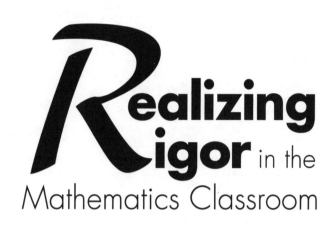

Realizing Rigor in the
Mathematics Classroom

Realizing Rigor in the Mathematics Classroom

Ted H. Hull
Ruth Harbin Miles
Don S. Balka

CORWIN
A SAGE Company

CORWIN
A SAGE Company

FOR INFORMATION:

Corwin
A SAGE Company
2455 Teller Road
Thousand Oaks, California 91320
(800) 233-9936
www.corwin.com

SAGE Publications Ltd.
1 Oliver's Yard
55 City Road
London EC1Y 1SP
United Kingdom

SAGE Publications India Pvt. Ltd.
B 1/I 1 Mohan Cooperative Industrial Area
Mathura Road, New Delhi 110 044
India

SAGE Publications Asia-Pacific Pte. Ltd.
3 Church Street
#10-04 Samsung Hub
Singapore 049483

Acquisitions Editor: Robin Najar
Associate Editor: Desirée A. Bartlett
Editorial Assistant: Ariel Price
Production Editor: Stephanie Palermini
Copy Editor: Codi Bowman
Typesetter: C&M Digitals (P) Ltd.
Proofreader: Christine Dahlin
Indexer: Terri Corry
Cover Designer: Karine Hovsepian

Printed in the United States of America.

Library of Congress Cataloging-in-Publication Data

Hull, Ted H., author.

Realizing rigor in the mathematics classroom / Ted H. Hull, Ruth Ella Harbin Miles, Don S. Balka.

pages cm
Includes bibliographical references and index.

ISBN 978-1-4522-9960-0 (pbk. : alk. paper)

1. Mathematics—Study and teaching—Standards—United States. I. Miles, Ruth Ella Harbin, author. II. Balka, Don S., author. III. Title.

QA11.2.H85 2015
510.71—dc23   2013037998

This book is printed on acid-free paper.

Certified Chain of Custody
SUSTAINABLE FORESTRY INITIATIVE
Promoting Sustainable Forestry
www.sfiprogram.org
SFI-01268
SFI label applies to text stock

14 15 16 17 18 10 9 8 7 6 5 4 3 2 1

Contents

Foreword   xi

About the Authors   xiii

Introduction   1
    CCSS Content and Practices   2
    A Clue to Rigor   3
    Outline of the Book   4
    How to Use This Book   5

PART I   **The Foundation**   7

1. Understanding and Meeting the Challenge of Rigor   8
    National Assessments   8
    Teacher Evaluation   9
    Learning Shifts   10
    Meeting the Challenges   10
    Looking at Assessments   11
    Rigor as a Common Factor   13

2. Defining and Instituting Rigor   14
    Searching for Evidence   15
    Dictionary and Thesaurus   16
    Professional Opinions   16
        Powerful Mathematics Instruction   17
        Rigor and Relevance   17
        Depth of Knowledge   18
    Indicators of Rigor   18
    Drawing Conclusions   19
    Decision Point   20
    Contrasting Lessons Examples   23
        Problem 1   23
        Problem 2   23
    Problem Analysis   23
        Problem 1   23
        Problem 2   24
        Problem Usage   25
    Transforming Classrooms to Support Rigor   26
    Having Productive Conversations   26

3. Building Team Leadership to Support Rigor   **27**
Role of a Steering Committee   29
Role of the Leadership Team   30
Role of the Principal   30
Developing Learning Communities   31
A Principal's Story   31
Having Productive Conversations   32

4. Rigor and the Standards for Practice   **33**
Standards for Mathematical Practice   33
Practice 1a—Make Sense of Problems   34
    Defining the Practice   34
    Recognizing the Practice in Action   34
Practice 1b—Persevere in Solving Them   35
    Defining the Practice   35
    Recognizing the Practice in Action   35
Practice 2—Reason Abstractly and Quantitatively   36
    Defining the Practice   36
    Recognizing the Practice in Action   37
Practice 3—Construct Viable Arguments and
  Critique the Reasoning of Others   37
Practice 3a—Construct Viable Arguments   37
    Defining the Practice   38
    Recognizing the Practice in Action   38
Practice 3b—Critique the Reasoning of Others   38
    Defining the Practice   39
    Recognizing the Practice in Action   39
Practice 4—Model With Mathematics   39
    Defining the Practice   39
    Recognizing the Practice in Action   40
Practice 5—Use Appropriate Tools Strategically   40
    Defining the Practice   40
    Recognizing the Practice in Action   41
Practice 6—Attend to Precision   41
    Defining the Practice   41
    Recognizing the Practice in Action   42
Practice 7—Look for and Make Use of Structure   42
    Defining the Practice   42
    Recognizing the Practice in Action   43
Practice 8—Look for and Express Regularity in
  Repeated Reasoning   43
    Defining the Practice   43
    Recognizing the Practice in Action   44
Rigor and Practices   44

A Principal's Story (Continued)   45
Having Productive Conversations   45

5. Rigor Related to Classroom Formative Assessment   **46**
Assessment Types   47
Classroom Formative Assessment:
   The Missing Instructional Element   48
Refining Formative Assessment   49
Classroom Formative Assessment   50
Formative Assessment and Intervention   51
Current Learning   51
Effective Intervention   52
Instructional Research   54
Synergy   56
A Principal's Story (Continued)   57
Having Productive Conversations   58

6. Rigor and the Proficiency Matrix   **59**
Organization   59
   Initial   60
   Intermediate   60
   Advanced   60
Progress Toward Rigor   60
Strategy Relationship in the Matrix   61
Classroom Formative Assessment and the Matrix   62
   Lesson Example: Using the Matrix to Select Strategies
      and Student Actions While Planning 62
   Domain: Geometry   66
   Problem   66
A Principal's Story (Continued)   67
Having Productive Conversations   67

PART II   Issues and Obstacles   **69**

7. Issues to Resolve   **70**
Issue: Teaching the Identified Content   70
Issue: Deepening Mathematical Understandings   71
   Making Connections   71
   Creating Meaning   72
   Using Learning Research   72
Issue: Reaching All Students   73
   Learning Opportunities   73
   Using the Strategy Sequence to Address Issues   76
   Using the Matrix   79
Having Productive Conversations   79

8. Obstacles to Success **80**

Obstacle: Working in Isolation 80

Obstacle: Attempting to Evaluate People to Change 81

Obstacle: Failing to Monitor Student Actions 82

Obstacle: Overadoption 82

Obstacle: Mistaken Efforts 82

Mathematics Adoption Analysis Tool 83

Understanding MAAT 87

Having Productive Conversations 87

PART III   Solutions **89**

**9. Solution Step 1: Monitoring Student Actions Related
to the Practices** **90**

Opening Classroom Doors 90

Nonevaluative Monitoring 91

Starting With Students 91

    Classroom Visit Types 92

    Classroom Visit Tally–Students 93

Teacher Self-Assessment of Student Actions 93

Math Coach Scenario 98

Having Productive Conversations 100

**10. Solution Step 2: Using Classroom Visit Data—Assessment
of Student Actions** **101**

Conducting Productive Conversations 101

Understanding Change Process 103

Levels of Adoption 104

Intervention as Support 105

Building a Critical Mass 105

Changing the Culture 106

Connecting Actions Chart 106

Math Coach Scenario (Continued) 108

    Meeting 2 110

Having Productive Conversations 111

**11. Solution Step 3: Monitoring Teacher Actions Related
to the Practices** **112**

Using the Classroom Visit Tally–Teachers Form 114

Conversations About the Data 115

Working on Individual Needs 115

Experimenting, Using, Integrating 119

Mathematics Collaborative Log 119

Teacher Planning Guide 119

Having Productive Conversations 119

12. Solution Step 4: Gathering and Using Additional Data   **122**

    Assessments Collectively   122

       Achievement Data   122

       Data From Classrooms   123

    Specified Classroom Visits   123

    Validity Visits   124

    Reverse Visits   124

    Teacher-Requested Visits   125

    Supporting Teachers' Change Efforts   125

       Experimenting   125

       Using   126

       Integrating   126

    Adoption Stages   126

    Documenting Progress   127

    Completing the Form   127

    Having Productive Conversations   131

13. Solution Step 5: Maintaining Progress Toward Rigor   **132**

    Background   132

    Relating Mathematical Rigor and the Practices   133

    Inferences From the Standards for

       Mathematical Practice   134

       Inferences on Content   135

       Inferences on Instruction   135

       Inferences on Assessment   136

       Inferences on Climate   137

    Rigor as an Outcome   137

    Categories   138

       Content   138

       Instruction   138

       Assessment   138

       Climate   138

    Rigor Analysis Form   139

    Explanation   139

    Directions   139

    Guiding the Work   142

    Having Productive Conversations   142

PART IV   Inputs and Outcomes   **147**

14. Teaching for Rigor   **148**

    Inputs   148

       Curriculum   148

       Classrooms   148

Outcomes 149
   Communication 149
   Culture 150
Teaching for Progress in Rigor 150
Having Productive Conversations 150

15. Coaching for Rigor **151**
Inputs 151
   Curriculum 151
   Classrooms 152
Outcomes 153
   Communication 153
   Culture 153
Coaching for Progress in Rigor 154
Having Productive Conversations 154

16. Leading for Rigor **155**
Inputs 155
   Curriculum 155
   Classroom 156
Outcomes 156
   Communication 156
   Culture 157
Leading for Progress in Rigor 157
Having Productive Conversations 158

PART V   Momentum **159**

17. Linking Responsibilities–Assessing Progress **160**
Professional Trust 161
Professional Conversations 161
Supporting Teacher Change 162
   Working to Improve 162
Documenting Change 163
   Using the Form 163
Conclusion 163
Having Productive Conversations 165

Appendix A. Standards of Student Practice in Mathematics Proficiency Matrix **166**

Appendix B. Instructional Implementation Sequence **168**

References **170**

Index **173**

Foreword

The Common Core State Standards for Mathematics brought "rigor" to the forefront of mathematics learning. Many educators do not understand the word rigor in an educational setting. New trends have emerged that demand students are taught and learn rigorous mathematics. The technological advances that continue to emerge demand mathematical rigor for all students in every classroom. Both teachers and leaders need to begin transforming classroom instruction, based on content knowledge and the Standards for Mathematical Practice, with the goal of implementing mathematical rigor in and out of the classroom.

Rigor can be embraced at all learning levels whether it is students, teachers, or leaders that are learning. Rigor requires a deep understanding of mathematics as well as the ability to transfer learning into new and challenging situations. Since there is no common understanding for rigor, teachers and leaders have a difficult task gauging whether or not mathematical rigor is occurring in classrooms or is consistently applied among classrooms.

The framework the authors of this book provide for analyzing how rigor is valued in the classroom beautifully captures the major problems and successes for learning mathematics. Rigor is a habit of mind and as such requires daily attention in the process of mathematics learning. Thinking, reasoning, and understanding of mathematics are all a basis of rigor. The authors believe that there are three widely agreed-upon principles that have demonstrated the nature of rigorous learning experiences:

1. good teaching is central to improving achievement;
2. teachers must identify rigorous, well-defined curriculum standards, benchmarks, and corresponding assessments; and
3. all stakeholders must hold high expectations for student performance.

The authors succeed in exploring and relating to rigor by applying the principles for helping students learn mathematics more deeply. They make the case for powerful mathematics instruction that relates conceptual development and mathematical connections. This type of instruction supports themes identified by the National Research Council as ones that support rigor. The authors also develop a Rigor Comparison Chart that lists current descriptions and thoughts about rigor versus future demands of rigorous mathematical teaching and learning. It is noted many times that mathematical rigor can be achieved by effectively implementing the Common Core State Standards for Mathematics and the Standards for Mathematical Practice.

Evidence of mathematical rigor is a direct result of active student participation in deep mathematical thinking and intensive reasoning. The degree of mathematical rigor is determined by the knowledge and understanding attained by every student.

The book explores ways teachers can take common problems and raise them to higher levels of thinking, moving in particular from arithmetic fluency to algebraic fluency and abstract thinking. In addition, leaders play a huge role in providing support for rigorous instruction which includes creating leadership teams, developing in-house professional learning communities, and providing access to professional learning from a variety of local, state, and national providers. Ultimately, change at the building level resides within the domain of the principal. Evidence of rigorous mathematical understanding from a variety of formative and summative assessments is essential to monitor student learning. The student Proficiency Matrix included in the reading can be used as evidence of student learning and can encourage teachers to apply the Standards for Mathematical Practice as students make progress toward rigor.

Each chapter has a set of questions called "Having Productive Conversations." These questions can extend the learning from each chapter by providing a way for leaders to guide conversations about rigor around content knowledge, the Standards for Mathematical Practice, and applications of the mathematics to challenging situations.

The authors offer a five-step approach to observing rigor in mathematics learning. The solution to rigor includes: 1) monitoring student actions related to the mathematical practices, 2) using classroom visit data to assess student actions, 3) monitoring teacher actions related to the mathematical practices, 4) gathering and using additional data, and 5) monitoring progress toward rigor. The strong ending to this book includes proven and researched inputs and outcomes that add to support for rigor. Necessary components to support rigor include teaching for rigor, coaching for rigor, and leading for rigor.

The value of this book is in its capacity to explain rigor in the context of teaching and learning mathematics. Teaching rigorously is a complex, demanding intellectual activity that requires the teacher to understand interactions between the content knowledge needed, the student-centered instruction required, and the student actions related to the Standards for Mathematical Practice. The authors have succeeded in presenting the case for rigor by developing definitions and tools that can be used to find evidence of student learning including a deep understanding of mathematics as well as the ability to transfer learning into new and challenging situations.

About the Authors

Ted H. Hull, EdD, completed 32 years of service in public education before retiring and opening Hull Educational Consulting. He served as a mathematics teacher, K–12 mathematics coordinator, middle school principal, director of curriculum and instruction, and a project director for the Charles A. Dana Center at the University of Texas in Austin. While at the University of Texas, 2001 to 2005, he directed the research project "Transforming Schools: Moving From Low-Achieving to High-Performing Learning Communities." As part of the project, Hull worked directly with district leaders, school administrators, and teachers in Arkansas, Oklahoma, Louisiana, and Texas to develop instructional leadership skills and implement effective mathematics instruction. Hull is a regular presenter at local, state, and national meetings. He has written numerous articles for the National Council of Supervisors of Mathematics (NCSM) Newsletter, including "Understanding the Six Steps of Implementation: Engagement by an Internal or External Facilitator" (2005) and "Leadership Equity: Moving Professional Development Into the Classroom" (2005), as well as "Manager to Instructional Leader" (2007) for the NCSM *Journal of Mathematics Education Leadership*. He has been published in *Texas Mathematics Teacher* (2006)—"Teacher Input Into Classroom Visits: Customized Classroom Visit Form." Hull was also a contributing author for publications from the Charles A. Dana Center: *Mathematics Standards in the Classroom: Resources for Grades 6–8* (2002) and *Middle School Mathematics Assessments: Proportional Reasoning* (2004). He is an active member of the Texas Association of Supervisors of Mathematics (TASM) and served on the NCSM Board of Directors as regional director for Southern 2.

Ruth Harbin Miles coaches rural, suburban, and inner-city school mathematics teachers. Her professional experience includes coordinating the K–12 Mathematics Teaching and Learning Program for the Olathe, Kansas, Public Schools for more than 25 years; teaching mathematics methods courses at Virginia's Mary Baldwin College and Ottawa, MidAmerica Nazarene, St. Mary's, and Fort Hays State universities in Kansas; and serving as president of the Kansas Association of Teachers of Mathematics. She represented eight Midwestern states on the Board of Directors for the NCSM and has been a copresenter for NCSM's Leadership Professional Development National Conferences. Miles is the coauthor of *Walkway to the Future: How to Implement the NCTM Standards* (Jansen Publications, 1996) and is one of the writers for NCSM's *PRIME Leadership Framework* (Solution Tree Publishers, 2008). As co-owner of Happy Mountain Learning, she specializes in developing teachers' content knowledge and strategies for engaging students to achieve high standards in mathematics.

 **Don S. Balka**, PhD, is a noted mathematics educator who has presented more than 2,000 workshops on the use of math manipulatives with PK–12 students at national and regional conferences of the National Council of Teachers of Mathematics and at inservice trainings in school districts throughout the United States and the world. He is professor emeritus in the Mathematics Department at Saint Mary's College, Notre Dame, Indiana. He is the author or coauthor of numerous books for K–12 teachers, including *Developing Algebraic Thinking with Number Tiles*, *Hands-On Math and Literature with Math Start*, *Exploring Geometry with Geofix*, *Working with Algebra Tiles*, and *Mathematics with Unifix Cubes*. Balka is also a coauthor on the Macmillan K–5 series *Math Connects* and coauthor with Ted Hull and Ruth Harbin Miles on four books published by Corwin. He has served as a director of the National Council of Teachers of Mathematics and the National Council of Supervisors of Mathematics. In addition, he is president of TODOS: Mathematics for All and president of the School Science and Mathematics Association.

Introduction

The Common Core State Standards are currently the driving force in educational change. Contained within the content standards are the Standards for Mathematical Practice. These Practices describe the varieties of expertise and abilities that mathematics educators should seek to develop in their students at all levels. These Mathematical Practices, if implemented, will significantly change student learning in mathematics by dramatically changing how teachers teach. Yet teachers are already feeling taxed to meet current demands.

Teachers face innumerable daily challenges in the course of fulfilling their jobs. Along with these daily challenges, teachers are also confronted by broader challenges that ebb and flow with societal and political trends. On this front, teachers are continually facing the "challenge de jour" approach to education. Rarely, it seems, are these types of challenges ever actually resolved. They merely become passé and are replaced by newer, more pressing challenges. Then, as is often the case, the original challenges resurface with new trendy terms. As a result of this tidal approach to reform, classroom instructional change has remained essentially inert.

In spite of this past history, some recent events have occurred that will unavoidably impact teachers and teacher leaders. With the adoption of the Common Core State Standards, a renewed focus has emerged that directly relates to student mathematics proficiency. The Common Core Standards, while impacting the nation, are state driven. States are devoting time, energy, and money into adopting and implementing the identified content, and more significantly, the Standards for Mathematical Practice. This bipartisan effort to improve mathematical learning is not going to fade away.

Leaders are being bombarded with terms and initiatives that end up pulling them in multiple directions, thus, creating lack of focus and clarity. Within this book, we show how all of the efforts directed at improving learning can be united, and when teachers and leaders work to implement the Practices, they are also accomplishing classroom formative assessment and providing for diversity and rigor. This united approach ensures all students are successfully learning mathematics and prepares them to demonstrate learning of mathematics on the upcoming PARRC and Smarter Balanced assessments.

CCSS Content and Practices

To encourage and support this change in instruction, teachers and leaders must understand the Practices. Following is a list of the Common Core Standards for Mathematical Practice. We provide commentary on them throughout the book.

CCS Standards for Mathematical Practice

1. Make sense of problems and persevere in solving them.
2. Reason abstractly and quantitatively.
3. Construct viable arguments and critique the reasoning of others.
4. Model with mathematics.
5. Use appropriate tools strategically.
6. Attend to precision.
7. Look for and make use of structure.
8. Look for and express regularity in repeated reasoning.

Leaders and teachers need to know what they look like when incorporated into classroom instruction. This is especially true for student actions, since the primary focus for change in the last decades has been on teachers' actions and teaching methods. With this shift from teacher focus to student focus, both teachers and leaders need assistance in reaching the goals of the CCSS. Moreover, teachers and leaders need to recognize the need for collaboration and to know how to effectively collaborate while working to implement the Practices. They need to share teaching strategies and lessons. What works for a particular practice? How much time is involved? How do we know what students are learning?

In the wake of Common Core, three important new trends have emerged. These trends, in conjunction with the Standards for Mathematical Practices, will challenge current teaching and learning processes and will fundamentally impact how teachers teach. These emerging trends all involve expanded use of technological advances:

1. *Assessing students' knowledge* beyond multiple-choice formats.
2. *Assessing students' mathematical proficiency, thinking, and reasoning.*
3. *Tracking students' knowledge gain* over extended periods of time.

These trends demand that students are taught, and learn, rigorous mathematics, a phrase that we detail in Chapters 2 and 3. Teachers and leaders must step forward to take ownership of the potential impact of these trends to meet the demands of mathematical rigor.

The technological advances demand mathematical rigor for all students in every classroom. Students' learning expectations in mathematics are expanding well beyond computational fluency because assessments are no longer bound to the constraints of multiple-choice test formats. The advances in assessment are happening now, as we have noted, and will continue. Teachers and leaders must be proactive by requiring students to think and reason mathematically to perform well on these challenging, open-ended assessments. Both teachers and leaders need to begin transforming classroom instruction with the goal of implementing rigor.

To date, however, mathematical rigor is rather loosely defined. While teachers have been pressed to become more rigorous, they have not been told what is actually expected as a result of being more rigorous. Rigor has been an elusive, unclarified expectation. Or worse, rigor has been narrowly defined and focused.

A Clue to Rigor

In our previous book, *The Common Core Mathematics Standards: Transforming Practice Through Team Leadership* (Hull, Harbin Miles, & Balka, 2011) we developed an easy-to-use Proficiency Matrix that is widely distributed and used. The Matrix lays out the Standards of Mathematical Practices called for in the CCSS and the expected outcomes for each level of student proficiency. We provide one row of the Matrix showing the three levels. In addition, strategies for implementing the Practices are provided in each cell (Grouping/Engaging, Encourage Reasoning).

	Students:	(I) = Initial	(IN) = Intermediate	(A) = Advanced
2	**Reason abstractly and quantitatively**	Reason with models or pictorial representations to solve problems. *(Grouping/ Engaging)*	Are able to translate situations into symbols for solving problems. *(Grouping/ Engaging)*	Convert situations into symbols to appropriately solve problems as well as convert symbols into meaningful situations. *(Encourage Reasoning)*

Teaching the CCSS content by incorporating the Practices as indicated in the proficiency levels of the Matrix makes it easier to achieve mathematical rigor.

The Matrix is a tool that was developed to assist teachers and leaders in effectively dealing with the necessary changes in teaching practices required by the Standards for Mathematical Practice and to steer their progress toward mathematically rigorous classrooms. It is designed to do the following:

- ❱ Focus collaborative conversations on implementation efforts.
- ❱ Serve as a guide for building effective lessons.
- ❱ Promote the Strategy Sequence of improving mathematical proficiency.

The Proficiency Matrix can play a critical role in helping teachers and leaders purposefully transforming classrooms to become more rigorous. The transformation to rigor is far more than just identifying content. The Practices focus learning related to students' proficiency, thinking, reasoning, and depth of understanding. The Proficiency Matrix helps teachers identify learning progression in each mathematical practice. The Matrix, combined with formative assessment strategies and techniques, helps teachers assess where students are in their learning progression. With this knowledge, teachers can adjust their teaching to achieve advanced proficiency of the individual practice and help students succeed.

This book, *Realizing Rigor in the Mathematics Classroom,* focuses directly on effectively implementing the Standards for Mathematical Practice. Using the Matrix as a framework, we will show you how achieving mathematical rigor is the same journey as is incorporating the Practices. We provide a step-by-step guide for leaders and teachers to use, as well as supporting tools to assist in charting progress. We identify actions that indicate how a practice in the classroom begins at one level and then how that same practice deepens over time as both students and teachers gain proficiency.

If ever an opportunity existed to change mathematics instruction to address the learning needs of every student, this is that time.

Outline of the Book

The first section of the book lays the foundation of the impact and trends associated with the implementation of the Common Core. It also discusses and defines rigor. As a result of the Standards for Mathematical

Practice, student learning expectations in mathematics are expanding beyond computational fluency and moving more toward student thinking and reasoning. As such, technological advancements in assessing students' knowledge, assessing students' mathematical proficiency, thinking and reasoning and tracking students' knowledge gain over time require a change in teaching practices. The Proficiency Matrix will be introduced at this point as a central tool to deal with implementation efforts.

The second section of the book focuses on potential issues or obstacles that will thwart teachers' and leaders' efforts to address the Common Core Standards for Mathematical Practices and increase student achievement. These include such things as differentiating instruction, monitoring classrooms, and using data. The entire book provides specific recommendations for meeting the challenges of the CCSS and effectively addressing the issues or obstacles.

The third section of the book provides a five-step solution process for implementing and sustaining the Standards for Practice and reaching the goal of providing a rigorous mathematics class.

The fourth section specifically addresses the necessary roles for teachers, coaches, and leaders to take to achieve implementation and rigor.

The last part of the book looks at how to sustain momentum for the implementation of the CCSS Mathematical Practices. It focuses on monitoring the many responsibilities of teachers, coaches, and school leaders to ensure successful implementation.

How to Use This Book

Because rigor is an outcome that is achieved by attending to specific inputs, attaining rigor takes time and focus. Mathematical rigor cannot be reached by maintaining our current instructional habits. Classroom instruction must incrementally change so that the climate and culture support rigor. Instructional change hinges on student thinking being made visible in every mathematics classroom by engaging students in instructional activities that require them to verbalize or demonstrate their understanding. The use of formative assessment information gathered from students through questioning, listening, or observing will help to immediately affirm or correct the displayed students' understanding.

To implement the changes necessary to achieve the goals of the Practices and mathematical rigor, teachers and leaders should read and discuss this book to first understand formative assessment in the context of mathematics. Once that is complete, they should begin to initiate use of the Proficiency Matrix and its recommended instructional strategies. Teachers and leaders should strive to undertake a cyclic pattern of implementation, monitoring, feedback, support, and improvement.

We provide numerous tools in the book for teachers and leaders to use to implement, monitor, refine, and sustain the Common Core Standards for Mathematical Practice. We have focused the book on productive conversations about student actions that directly relate to achieving the Practices, and the formative assessments necessary to impact change.

We emphasize students and the actions students should "engage" with mathematics. Any educator wishing to implement the Standards for Practice in a way that produces mathematical rigor will find our book beneficial, practical, and easy to use.

PART I

The Foundation

Leaders and leadership teams need to build a solid foundation on which they can foster and support change initiatives that promote student learning. The foundation consists of recognizing some pressing issues, building teams of leaders, gaining critical knowledge of formative assessment, understanding the Practices, and studying the Proficiency Matrix. These are the essential ingredients needed to purposefully institute change in instructional practices that support rigor.

The foundation also requires a sense of urgency for why change is necessary. A sense of urgency alone is not sufficient to motivate and sustain change efforts. However, a sense of urgency is sufficient and necessary to begin initiating conversations and promoting a need to plan for action.

1

Understanding and
Meeting the Challenge of Rigor

We are a nation consumed and driven by testing. There seems to be a test for every condition, desire, personality, and career choice. We seek to explain and identify who we are, and what we should do or become, by taking tests. Tests, frequently multiple-choice, are used to determine our capacity to perform in certain arenas of life. Clearly, schools are no exception to the test-taking phenomenon and are, perhaps, more consumed by testing than many other systems or institutions.

We are also a nation in a hurry. Time always seems to be a pressure point. In response to this apparent pressure, we expect tests to be rather brief in nature with short responses, and quick to score. While there is no inherent problem in administering or taking tests, difficulties arise when tests and their results are either not considered or are used as the only source to make decisions. These situations are especially problematic when the decisions made directly impact or control an individual's life. Issues now arise as to whether the test is reliable, valid, and unbiased. Is the very fine line between a score of 68 and 70 truly accurate enough to determine the placement of an individual adult or child? There is rarely something definitive derived from a single test.

National Assessments

Regardless of concerns that may be raised around testing, and the accuracy of the resulting decisions, testing for students is solidly embedded from the national level down. Individual states, individual districts, individual schools, and individual students are all defined by test results. The results can lead to praise and acclaim, or they can be devastating. Testing programs and test results greatly influence school decisions and actions at all levels. Testing is a reality that shows no signs of abating.

The Common Core State Standards are no exception to the pressures of testing either. Once the content is adopted into the individual state standards, the content is formatted into tests. Students are tested to determine their degree of mastery of the content. Yet there are issues being raised.

Within the Common Core Content Standards are the Standards for Mathematical Practice. These Standards are being widely distributed and the pros and cons are being discussed. Leaders and teachers are attempting to relate the Practices to instructional research. In July 2013, Presidents of the Conference Board of the Mathematical Societies (Conference Board of the Mathematical Sciences, 2013) put forth a statement of support for the Standards, noting, "If properly implemented, these rigorous new standards hold the promise of elevating the mathematical knowledge and skill of every young American to levels competitive with the best in the world, of preparing our college entrants to undertake advanced work in the mathematical sciences, and of readying the next generation for the jobs their world will demand." Furthermore, teaching strategies will need to shift to meet the demands of the Practices. For this reason, serious work is taking place to provide tests that actually assess students' conceptual understanding, thinking, and reasoning in mathematics.

Since curricular decisions are strongly influenced by state-administered tests as well as district-level ones, there are definite reasons to believe that this shift in assessment will also greatly influence curricular decisions. There is, nonetheless, a caveat. In general, state and district tests have greatly influenced the mathematics content that is taught. This testing impact has not been nearly as significant on instructional strategies. Now, with the Practices being assessed, instructional strategies must change if students are to perform even satisfactorily on the newly developed tests by Partnership for Assessment of Readiness for College and Careers (PARCC), Smarter Balanced, or individual states. If students are truly assessed on their thinking, reasoning, and problem-solving abilities, then they must spend significant time thinking, reasoning, and solving challenging problems.

Teacher Evaluation

The demand to shift instructional strategies to increase students' mathematical learning is rather intense by itself. Focused conversations and, hopefully, purposeful professional learning opportunities are occurring around the Standards for Mathematical Practice and the related instructional strategies that support the Practices in the classroom.

However, this is not the only demand that is surfacing. On a parallel path, and not tied to Common Core adoption, is a significant national trend concerning teacher performance. There is a tremendous push by a variety of stakeholders to directly relate teacher evaluation to student performance. This push is not about a general, nonspecific relationship between overall teachers' performance and students' performance, but rather specifically tied to a teacher and his or her students' progress, with students' progress being heavily, if not solely, defined by a test.

Teachers, understandably, have legitimate concerns about relating student performance with student assessment results. The relationship is difficult to accurately demonstrate using current testing techniques. Multiple-choice tests derived from specific content are fraught with accuracy issues when taken to an individual student level. The problem is greatly increased when tests are offered as a onetime event at the end of a school year.

The difficulties exist even if the tests are statistically valid and reliable. Issues related to student learning and multiple-choice test scores are numerous, but what happens when future technological advances allow for a wider variety of assessment formats, and more accurate tracking of individual student progress? We all shall quickly see because that future is here as PARCC, Smarter Balanced, and states continue to roll out various sample test items.

Learning Shifts

Teachers are faced with some very difficult and serious questions that cannot be ignored. With the trend of evaluating students at a conceptual level of learning and understanding, and with students' test results directly impacting teachers' evaluations, teachers need to carefully consider how to address the challenges. Addressing the forthcoming challenges requires thinking about teaching and learning in a whole different perspective—from the student's point of view. One thing is for sure, maintaining the current instructional approach—focused on teacher actions and reactions—will not prove a successful way of meeting the rising challenges. Yet teachers must not feel overwhelmed. Manageable, successful shifts that we describe in this book are achievable.

Meeting the Challenges

Common Core adoption is moving forward. The Standards for Mathematical Practice are moving forward. Assessment shifts are moving forward, and technology to support assessments is moving forward. Teachers, mathematics leaders, and school leaders do not have time to waste; they too must move forward. Standing still and waiting to see what happens is just not a good decision. The signs are pointing in the direction that more challenging assessments are here.

While the issue of student performance and teacher evaluation will take many twists and turns, and appear in different states in varied ways, the issue shows no indication of fading. This issue, while certainly a concern, should not be the motivating force for teachers to change their instructional strategies to meet the demands of the Practices. The motivating force should be that there is only positive that comes from shifting instruction so that practically every student successfully learns mathematics. There is absolutely no downside to using the Practices to teach the Common Core content. When students are able to think and reason mathematically, and when they understand mathematical concepts and connections, they will excel on any form of assessment. Moreover, students greatly increase their chances of excelling in their selected career and open possibilities for more career paths when they understand mathematical concepts and connections. Attaining this type of mathematical rigor through incorporating the Practices is a win-win situation.

Looking at Assessments

To help understand what students are expected to be able to do on forthcoming assessments, some examples may prove helpful. These examples, while significant, do not display all the different ways technology allows for answer choices to be recorded. New technology provides answers that can be "dragged and dropped" into an answer format. Free responses may be "bubbled in" or recorded by hand. In some cases, where multiple choices are provided, more than one answer is correct, rather than just one of the A, B, C, or D choices. New assessment items will be dynamic and interactive. Two problem examples are provided in Box 1.1.

After reviewing these two problems, it is obvious that they are more challenging for students, and they are also the very types of problems our students should do. A "traditional" item similar to Example 2 might have shown the first figure and merely asked students to shade 1/6 of the figure. With the new wording of the problem, students must exhibit some type of spatial sense, an understanding of diagonals of a regular hexagon, possibly an understanding of perpendicular bisectors of a line segment, and the identification of equilateral triangles or kites. Newer

Box 1.1

Sample Problem 1

The five fastest recorded times without wind assistance for boys under the age of 18 in the 100-meter dash are the following: 10.19, 10.23, 10.24, 10.25, and 10.26.

If the five boys ran a race, explain how the results of the race would change if the timers used stopwatches that rounded to the nearest tenth.

(Statistical times from: wikipedia.org/wiki/100_metres#Youth_.28under_ 18.29_boys)

Based on the times to the nearest hundredth, 10.19 is the fastest time. However, when rounded to the nearest tenth, the times become 10.2, 10.2, 10.2, 10.3, and 10.3, creating a three-way tie for first place.

Sample Problem 2

Mariana is learning about fractions.

Show how she can divide this hexagon into six equal pieces. Write a fraction that shows how much of the hexagon each piece represents.

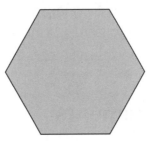

(from http://www.ccsstoolbox.com/parcc/PARCCPrototype_main.html and http://www.parcconline.org)

Explanation: As noted by PARCC, this particular Grade 3 sample item addresses more than one content area of mathematics. In this case, Number and Operations—Fractions, Measurement and Data, and Geometry are all involved. There are at least two ways to partition the hexagon into six equal pieces with each piece having a value of 1/6. Two common solutions are shown here. In the first figure, students understand that the diagonals of a hexagon partition it into six equilateral triangles. Therefore, each equilateral triangle has a value of 1/6. In the second figure, bisectors of the parallel sides of the hexagon are constructed, creating six congruent kites. Each has an area of 1/6.

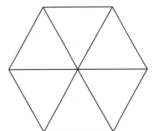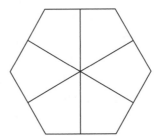

assessment items better indicate student understanding, and serve to resolve many of the difficulties outlined at the beginning of the chapter concerning limitations of multiple-choice tests.

Rigor as a Common Factor

The elements we have described—teacher evaluation shifts, assessment shifts, learning shifts—all have something in common. They demand that our students be engaged in a rigorous mathematics program. Rigor requires a deep understanding of mathematics, the type of understanding where students can transfer their learning to novel situations. This depth of understanding allows students to successfully meet new assessment challenges. Superficial exposure to skills will not lead to student success. Rigor, even though the term has been in the mathematics vocabulary for some time, has never been truly clarified. More important, rigor has not been defined in a way most educators commonly accept as accurate. Since there is no common acceptance, people use the term for their own purposes and with their own meaning. As a result, mathematics teachers and leaders have a difficult task gauging whether mathematical rigor is occurring in classrooms or if it is consistently being applied from classroom to classroom. However, with the Common Core content and Practices, that task is about to change. Rigor, then, must be explored and clarified.

2

Defining and Instituting Rigor

Mathematics teachers in classrooms across the United States are being pressured by federal and state departments of education to make learning of the content more rigorous. Rankings of U.S. students on international mathematics examinations compared to students in other countries have been one of the major points that have caused the situation to grow. The response to this pressure has been to "speed up" rather than "delve down." Rigor has been assumed to mean that mathematics courses, or grade levels, must contain more content, that students be introduced to concepts at earlier grades, that procedural fluency include larger numbers, and that, essentially, students complete more work. What is perceived to be rigor, within this context, increases the learning gap, thus, creating inequity. Furthermore, with few exceptions for a very small minority of students, these approaches actually undermine rigor.

Fortunately, there is a much better alternative meaning for rigor. Rigor is a habit of mind. Thinking, reasoning, and understanding are the basis of rigor. Students and teachers cease "skimming the top of mathematics" and actually explore in more depth the concepts appropriate to the age of the student. Students learn to logically construct connections between mathematics concepts. Students solve challenging problems and then articulate their thinking about the processes and their conclusions.

When studying and aligning the content within the Common Core, leaders may feel pressured to make curriculum documents appear more "rigorous" by further pushing the content into lower grades or using pacing guides that encourage less depth of study. This is a serious error that will undermine what is intended by the CCSS and the Standards for Mathematical Practice.

According to Hall and Hord (2001, p. 4), "Change is a process, not an event." Instituting mathematical rigor as we describe requires significant change occurring over time and is, therefore, a process, not an event.

This realization is a "must get" for mathematical rigor to be attained. Mathematical rigor is a result, a goal. Leaders and teachers do not begin with rigor; they reach the goal of rigor by carefully implementing the required foundational elements that lead to mathematical rigor. This message is so important that it is revisited throughout the book, and it must be supported by evidence.

Searching for Evidence

The National Research Council (NRC) is highly regarded and respected for its contributions to mathematical research on student learning and achievement. In trying to clarify and define mathematical rigor, NRC is a logical place to start. In reviewing the glossary for *Adding It Up* (2001), the discovery is quickly made that rigor is not present in the listing. *Adding It Up* does not specifically reference rigor, yet the document identifies significant ideas related to rigor in the mathematical strands referenced by the Common Core writers in relation to the Standards for Mathematical Practice. NRC states

Mathematical proficiency, as we see it, has five components, or strands:

- Conceptual understanding—comprehension of mathematical concepts, operations, and relations
- Procedural fluency—skill in carrying out procedures flexibly, accurately, efficiently, and appropriately
- Strategic competence—ability to formulate, represent, and solve mathematical problems
- Adaptive reasoning—capacity for logical thought, reflection, explanation, and justification
- Productive disposition—habitual inclination to see mathematics as sensible, useful, and worthwhile, coupled with a belief in diligence and one's own efficacy. (p. 116)

Reviewing the glossaries for *How Students Learn* (NRC, 2005), *How People Learn* (NRC, 2000), and *Engaging Schools* (NRC, 2004) produces the same absence of references to rigor.

At first, this may seem rather puzzling, surprising, and somewhat disconcerting. With all the emphasis on mathematical rigor, logic dictates that rigor has been carefully studied and defined. This is just not the situation. For this reason, the case for explaining and understanding rigor needs to be made. Because NRC does not specifically use the word "rigor," it does not mean they do not provide guidance to aid the search. Mathematical proficiency as defined previously certainly correlates to rigor. Additional research that points to indicators of rigor needs to be highlighted and the specific indicators identified. The search to define and explain rigor first begins with dictionary and thesaurus findings.

Dictionary and Thesaurus

A variety of words or phrases permeate the "rigor" domain; however, only a few actually lend themselves to discussing the notion of rigor in mathematics such as thoroughness, firmness, use of demanding standards, strict accuracy, complex thinking, depth of understanding, and severe exactitude. These meanings for rigor seem rather harsh, and leave mathematics educators in a quandary when responding to the often-heard question: What is a rigorous mathematics curriculum? In the paragraphs that follow, we highlight various aspects of the words described. Research on the topic is limited, but discussion is prevalent.

Professional Opinions

For example, state departments of education, such as Iowa, put forth their own ideas about rigor. The Iowa Department of Education (2005) states:

> Three widely agreed upon principles have been demonstrated to create rigorous learning experiences: (1) good teaching is central to improving achievement; (2) teachers must identify rigorous, well-defined curriculum standards, benchmarks, and corresponding assessments; and (3) all stakeholders must hold high expectations for student performance (p. 4).

While these three principles may be accurate, they only serve to perhaps point the direction and raise additional questions. What is "good" teaching? What is a rigorous, well-defined curriculum, and are teachers the primary source? Who are stakeholders, and what are high expectations? Understanding and attaining mathematical rigor requires much more clarification.

Almost every mathematics professional has some idea about rigor. Given a few minutes, and the opportunity to do so, professionals could explain what they believe rigor entails. Surprisingly, professional opinions do not abound in literature. Rigor, or the lack of, is frequently used during discussions concerning mathematics teaching and learning, but not so in printed text. Three sources are discussed that offer professional opinions about rigor.

Powerful Mathematics Instruction

In *Powerful Mathematics Instruction* (Pittsburgh Science of Learning Center, 2007), the authors state:

> During the last two decades, a diverse group of researchers and educators have developed and implemented approaches to instruction that reflect this consensus. These integrated instructional approaches generally combine three dimensions of teaching: *intensive use of classroom discussion, mathematical depth and rigor in the curriculum and in its implementation, and attention to student reasoning.* Many studies have yielded promising outcomes, including significant improvements in student learning among low income and minority students. (p. 1)

The quote identifies mathematical depth and rigor in the curriculum and during instruction. The quote also refers to "this consensus." The points identified in the consensus are explained in the forthcoming section titled Indicators of Rigor.

The authors of *Powerful Mathematics Instruction* (Pittsburgh Science of Learning Center, 2007) refer to rigor in several ways but highlight two specific characteristics. First, they reference teacher mathematical knowledge. There is a relationship (perhaps not causal due to many possible variables) between teachers' mathematical content knowledge and student achievement. The second reference to rigor addresses concept development. The authors state that students who receive instruction that builds conceptual understandings outperform those who do not receive such instruction. Rigor relates to conceptual development and is about fostering mathematical connections between facts and procedures and their underlying concepts.

Rigor and Relevance

Daggett (2005, p. 2) has developed a "Rigor/Relevance Framework" used to "aspire to teach students to high rigor and high relevance." The Framework is not mathematics specific. Using the Framework assists teachers in uniting curriculum, instruction, and assessment. The Framework is developed by using Bloom's Taxonomy for knowledge and cross-discipline applications to make the content relevant. Four quadrants are created. The quadrants are the following:

A—Acquisition
B—Application
C—Assimilation
D—Adaption

All four quadrants are used, each with a specific purpose, and shift from the teacher doing the work to the students doing the work. Essentially, A is teacher work, B is student work, C is student thinking, and D is student thinking and working (Daggett, 2005, p. 3). This quick recap is in no way intended to adequately explain the Rigor/Relevance Framework. The information is only to highlight that the Framework exists, and it is designed to increase rigor by having students engage in real-world, relevant problems.

Depth of Knowledge

Webb (2002) developed four depth-of-knowledge levels for analyzing content standards and assessment items: Level 1 is recall, Level 2 is Skill/Concept, Level 3 is Strategic Thinking, and Level 4 is Extended Thinking. His work proves useful in discussing and clarifying mathematical rigor. While the identified levels do not specifically identify when or if rigor is attained, the system does identify increased levels of student thinking and reasoning. If higher cognitive demands on students to think and reason are believed linked to rigor, then the depth-of-knowledge levels also indicate rigor. This assumption seems to be reinforced by the National Center for Research on Evaluation, Standards, and Student Testing (CRESST) when it uses Webb's Depth of Knowledge (DOK) scale to rate items in their report, "On the Road to Assessing Deeper Learning: The Status of Smarter Balanced and PARCC Assessment Consortia" (Herman & Linn, 2013). The report discusses the Center's monitoring efforts related to Smarter Balanced and PARCC's efforts to develop instruments that measure deeper learning.

Indicators of Rigor

In our quest to uncover the meaning of rigor, *Powerful Mathematics Instruction* (Pittsburgh Science of Learning Center, 2007) was quoted in the section titled "Professional Opinion." A "consensus" emerges that pinpoints similar ideas cited in previous paragraphs:

> Recognition of the need to integrate students' conceptual understanding, procedural competence and communication abilities is supported by 30 years of cognitive science research. This research shows that accurate knowledge is foundational for learning with understanding. At the same time, it is clear that long-term retention of factual material is best secured when learners come to understand the logic and organization underlying the facts. Furthermore, the research shows that learning is most robust when learners become actively engaged in reasoning about the knowledge they are acquiring (Bransford, Brown, & Cocking, 2000; Resnick & Hall, 1998, p. 1).

The elements identified in the passage are indeed prevalent in research. When reviewing the research about the abilities students need to have to learn mathematics, themes emerge. These themes strongly indicate mathematical rigor when rigor is deemed to be depth of understanding.

Returning to the National Research Council with their long history of focusing on learning, there are clearly identified themes that promote rigor. Some of the themes are the following:

▶ Knowledge transfer
▶ Efficient retrieval systems (logic and organization)
▶ Communication of ideas and understandings
▶ Perseverance (learning from previous efforts, and the ability to remain focused)
▶ Problem-solving strategies (thinking and reasoning)
▶ Skill proficiency
▶ Engagement
▶ Learning from others, self-monitoring, metacognition
▶ Justifying and explaining

The list could continue, but the point is made. If what is known from learning research is applied to classroom instructional practices, the course will be rigorous. Both professional opinions (Daggett, 2005; Pittsburgh Science of Learning Center, 2007) relate rigor to students being engaged in challenging tasks that require thinking and reasoning.

Drawing Conclusions

There is no perfect definition of rigor. There is agreement that deep understanding of mathematics is essential. The definition remains nebulous and imprecise, not only in mathematics but also in other content areas. Rigor is still being debated, and will continue to be debated for some time. Nonetheless, state and national assessments, by practically any definition, are becoming more rigorous. No longer are single-response or multiple-choice test items the main types of testing items. Students must explain solutions to a variety of problems. A problem scenario might be provided that requires in-depth understanding of technological advances in assessment, opening many pathways to what mathematical content is assessed and how the content is assessed. The overall goal is to better understand what students truly understand within the mathematics discipline. The assessments require skills but are not skill driven. They are problem-solving driven with thinking, reasoning, and conceptual knowledge being the keys to success.

This shift in learning expectations, driven by assessment expectations and capabilities, demands a shift in instructional strategies. The National Research Council (2012) is again guiding learning research. In identifying what NRC refers to as "21st-century skills," it has grouped competence beneath three domains: cognitive, interpersonal, and intrapersonal.

- ▶ Cognitive: thinking, reasoning, problem solving, and memory
- ▶ Interpersonal: express information to others, interpret others' messages, and appropriately respond
- ▶ Intrapersonal: emotions, feelings, and self-regulation (p. 21).

This grouping is designed to emphasize the importance of what NRC describes as "deeper learning." NRC (2012) states:

> As noted earlier, the three major reform documents in school mathematics— CESSM, PSSM, and CCSSM—all emphasize deeper learning of mathematics, learning with understanding, and the development of usable, applicable, transferable knowledge and skills. (p. 119)

The acronyms in the quote stand for the following:

> CESSM: Curriculum and Evaluation Standards for School Mathematics (NCTM, 1989)
> PSSM: Principles and Standards for School Mathematics (NCTM, 2000)
> CCSSM: Common Core State Standards for Mathematics (NGA, 2010)

These are important and influential documents. They also indicate an extremely high correlation between the research-based instructional strategies and the learning themes from NRC and the Standards for Practice as indicated in the Proficiency Matrix that we will describe and discuss.

Rigor, then, requires a shift in our thinking. Leaders and teachers need to understand some of the shifts in mathematical rigor requirements from current to future. Table 2.1 compares some of the common thoughts and beliefs about rigor to the rigor we believe the Practices support.

Decision Point

Teachers and leaders are at a critical decision point. They are confronted with the notion of rigor within the CCSS and the Practices. Consequently, they must formulate, articulate, and build agreement on what they believe mathematical rigor is. Many questions involving ideas that have previously been posed stand before them.

Table 2.1 Rigor Comparison Chart

Current Descriptions and Thoughts	Future Demands
• Material is difficult.	• Material is challenging.
• Pace is rapid.	• Pace is slower, but deeper. Lessons are scaffolded to push deeper.
• Students work independently.	• Students work collaboratively.
• More problems are completed.	• Fewer problems are completed, but more student work is done with understanding required.
• More homework is assigned.	• Homework is interesting and a natural extension.
• Teacher is sole information source.	• Entire classroom is information resource.
• Climate is tense, with a sense of pressure.	• Climate is supportive and encouraging.
• Content is accelerated.	• Content connections are stressed.
• Classrooms are appropriate for elite and gifted.	• Content is appropriate for student's ability. • Instruction is appropriate for all students.
• There are high expectations for keeping up.	• There are high expectations for success.
• Procedural fluency is critical.	• Thinking and reasoning are critical.
• Skills and computation are the focus.	• There is a concept focus with skills as tools for understanding.

Can mathematical rigor be achieved by

1. Only including additional content?
2. Only offering acceleration of courses?
3. Including an idea superimposed upon the mathematics program?
4. Effectively implementing the CCSSM Content Standards and Practices?
5. Doing something else entirely?

There is clearly no doubt, and it will come as no shock, that we promote and support Option 4.

While teachers greatly influence the climate and conditions within their classrooms, mathematical rigor within classrooms is actually what the students are doing to increase their learning. Rigor is a direct result of active participation in deep mathematical thinking and intensive reasoning. The degree of mathematical rigor is determined by the knowledge and understanding attained by every student.

With this information, we realize the need to offer dual meanings for rigor. First, we need to define mathematical content rigor, but then we also need to define mathematical instructional rigor. This leads us to our two definitions of rigor.

> **Content:** *Mathematical rigor is the depth of interconnecting concepts and the breadth of supporting skills students are expected to know and understand.*

Instructional rigor must be solidly based on challenging and worthwhile mathematical content. While content is essential, our focus in this book is on implementing the identified rigorous content through powerful instruction. For this reason, we offer a definition of instructional rigor.

> **Instruction:** *Mathematical rigor is the effective, ongoing interaction between teacher instruction and student reasoning and thinking about concepts, skills, and challenging tasks that results in a conscious, connected, and transferable body of valuable knowledge for every student.*

We believe mathematical rigor is a goal that can be achieved by effectively implementing the CCS Content Standards and Practices. To facilitate accomplishing Option 4, we have developed the Proficiency Matrix. The Matrix is a tool that identifies proficiency levels for each Standard for Practice. Our premise concerning the achievement of mathematical rigor is stated here.

> **Premise:** *Teaching the Common Core content using the Standards for Mathematical Practice to reach progressively higher levels of proficiency attains mathematical rigor.*

Leaders and teachers must trust and believe in this premise: The Standards for Mathematical Practices support visible thinking, ongoing formative assessment, and intervention. These three factors work together to achieve higher levels of proficiency and, thus, achieve mathematical rigor. As teachers and leaders strive to implement the factors into action in classrooms, rigor is being attained. The journey may not always be easy, but the results are well worth the effort.

Contrasting Lesson Examples

Often, a picture is a far better communicator than words. For a book on mathematics, the picture is a mathematical problem. One problem presents a traditional word problem used as either practice or assessment, and the contrasting problem (for learning and/or assessment) provides an opportunity for discussion. We borrow a problem from our own work (Hull, Balka, & Harbin Miles, 2012) because we have found the following comparisons to be easily identified regardless of a participant's background or interest.

Problem 1

If I have two pennies, a nickel, two dimes, and a quarter, how much money do I have?

A. 41¢   B. 45¢   C. 52¢*   D. 77¢

Problem 2

I have five coins in my pocket. The coins may only be pennies, nickels, dimes, or quarters. If I reach into my pocket and pull out three coins, how much money might I have in my hand?

Problem Analysis

Problem 1

In Problem 1, students need to recognize the word names for coins, translate the word names into coin values, and then add the values to determine the amount. While this is a skill students need to do, the process is at recall and knowledge levels. The brain does process the information, but to say that thinking and reasoning are required to any degree beyond the surface level is a real stretch of one's imagination.

To test this last statement, one might conjecture about certain questions to be asked that would push the students to more depth of understanding. Some questions may include the following:

- How did you get the answer?
- Describe what you were thinking while solving the problem?
- Who solved the problem differently?

Still, the sample questions do not focus on higher conceptual understanding. Each of these questions requires students to basically recall and state the procedures they followed in reaching an answer. Students will probably point out the procedure of writing amounts for the coins and then adding the amounts to reach a total. Student reasoning, as a part of rigor, is lacking or, at best, minimal. The questions do not provide any opportunities for students to reason abstractly. Perseverance is also minimal; there is a single correct answer.

Problem 2

The transformed coin problem offers a much better opportunity for analysis of student learning and understanding. Teachers have several entry points to the problem and then several pathways in which to explore learning more deeply. Next is one of the ways the problem may unfold. The problem is layered for depth of exploration and understanding. This layered approach provides a scaffold so all students can engage in the problem.

Part A

Teachers explore with students the different ways there could be five coins. With plastic money, students could select and discuss combinations of five coins, then share what they selected. Teachers are looking to see how students select and arrange possible combinations. Since there are four different coin types (penny, nickel, dime, and quarter), students must decide to duplicate one of the coins to have five of them.

Part B

After students have explored the five coins possibilities, teachers now push students to think about the combinations of three coins that could be in the student's hand. There are many possible combinations, and students are not expected to find them all. The probing questions asked are listed here:

- What three coins might you have in your hand? Explain.
- What is the least amount of money you could have in your hand? Explain how you know.
- What is the greatest amount of money you could have in your hand? Explain how you know.

Part C

After students have discussed possible answers and agreed that 3¢ is the least amount, students need to discuss and reach agreement on the greatest amount of money held in one's hand. As a note, students may justifiably argue that the least amount is 16¢ (penny, nickel, dime) since you may have to have different coins. This is logical, and teachers should return to the original stated problem and ask students to clarify and discuss their reasoning. Teachers may find it necessary to add clarifying statements to the original problem so everyone agrees that 3¢ is the answer since the problem uses plurals for coins.

Once agreement and clarity are reached, teachers ask the following:

> If 3¢ is the least amount of money you could have in your hand, what is the next least amount you could have in your hand? What coins would you have and how do you know this is the next least amount?

> If 75¢ is the greatest amount of money you could have in your hand, what is the next greatest amount of money you could have in your hand? What are the coins and how do you know this is the next greatest amount?

Part D

Starting with the least amount of money (3¢), discover and list the amounts of money, from least to greatest, you could have in your hand with three coins. After the exploration time, teachers want to ask probing questions such as the following:

▶ Do you have every answer? How do you know?
▶ How did you find the answer?
▶ Did you have an organization strategy? What did you do to get your answers?

Part E

What if you reached into your pocket and grabbed four coins instead of three? Can you use the information you have to arrange the possible coin selections and amounts?

Problem Usage

In Problem 2, teachers have several options when working with problems of this nature. First, it is not necessary for every problem part to be used. Second, not every student may do every part. Third, the parts are not necessarily done in one instructional period. Fourth, the problem can be revisited with more complex parts completed days or even weeks later. Fifth, teachers should feel free to follow students' discoveries and interests as the problem unfolds.

Now, consider the same three questions asked at the end of Problem 1:

- How did you get the answer?
- What you were thinking while solving the problem?
- Who solved the problem differently?

Student responses are limited for each question. However, in Problem 2, students have multiple paths to follow when working toward and reaching answers. Students must actually think, test, reflect, organize, clarify, and explain their solution path and possible answer. Finally, groups of students or an entire class can compare and contrast the various ways to approach, organize, and solve different parts of the problem. The classroom teacher serves as a facilitator in the discussion but also provides a metacognitive approach to the problem.

Transforming Classrooms to Support Rigor

These two problems exemplify the differences between current mathematics lessons and tests and lessons and assessments that support the development of mathematical rigor. The difference gap is rather vast and will not be bridged without focused time and energy. Collaboration is required at every level of the system—student, teacher, mathematics leader, and school leader. An effective collaboration process emerges in the ensuing chapters.

Having Productive Conversations

1. Are you being encouraged to provide more rigorous mathematics?
2. What was your thinking and understandings of rigor before reading this chapter?
3. What has changed about your thinking and understandings after reading and discussing this chapter?
4. What do you see as the relationship between rigor and the Standards for Practice?

3

Building Team Leadership to Support Rigor

The Common Core State Standards are the most exciting initiative introduced to education for decades. There are no initiatives that are even similar in scope. The focus on high expectations for every student with common content across the nation is filled with possibilities. In mathematics, the Standards for Mathematical Practice provide a clear message for how students should be engaged in learning content.

Obviously, if students are to learn mathematics differently, teachers must teach mathematics differently, and leaders must lead differently. Once again, a collaborative effort is required.

Leadership teams are very important for the change process. Teams formulate and internalize a common vision, establish clear expectations, get everyone involved, and focus on achieving the common established goals.

The responsibilities for teachers and leaders related to mathematics achievement are, and must be, cohesive and related. To have cohesive, interrelated responsibilities, all parties must work collaboratively toward common goals. Collaboration means there is clarity in communication, content, classroom actions, and climate conditions. Teachers and leaders need to individually and collectively understand their roles and responsibilities, yet work collaboratively to achieve them. This chapter assists leaders and teachers in forming collaborative teams.

Roles and responsibilities, while interrelated, are also unique. For this reason, there is a need to periodically identify specific responsibilities to specific positions. Throughout this book we use the terms leader, school leader, mathematics leader, and team leader. When the term "leader" is used, we mean anyone (including teachers) who is in a role to monitor and guide the direction of the initiative for change. The term "school leader" specifically relates to principals, assistant principals, and perhaps individuals from central office who evaluate personnel. The term "mathematics leader" is reserved for individuals whose content specialization is mathematics, and they operate in a support role. These individuals may be coaches, coordinators, supervisors, or specialists.

The Standards for Mathematical Practice spotlight what students must be doing to successfully learn mathematics content identified in the Common Core. The Practices are not done "to" students but are done "by" students, as the students actively engage in carefully constructed mathematics lessons. Each identified practice is rather brief in description but extensive in meaning. The Practices, when carefully studied, unfold in layers upon layers of understanding for various mathematical relationships. They also vary by grade level or subject. As a result, mathematics leaders and teachers must initially understand the overall purpose of the Practices and continually deepen their understanding over time.

Here is a simple example that we have used in presentations. It is part of a more extensive six-part problem. The directions have been modified for this particular case.

> Use the digits 3, 4, 5, and 6 to complete the number sentence. A digit can only be used once in the number sentence.

$$\underline{\quad}\,\underline{\quad} - \underline{\quad}\,\underline{\quad} = 31$$

Upon cursory view, the problem seems to involve only computational fluency with two-digit subtraction and some reasoning about basic facts. With little effort many students would find a solution quickly: $65 - 34 = 31$. There is important reasoning involved. What two digits can I use in the ones place to give a difference of 1? What two digits can I use in the tens place to give a difference of 3? Are there other possible solutions?

Teacher efforts through questioning now come into play. This problem involved four consecutive integers, 3, 4, 5, and 6. Could you get an answer of 31 with four different consecutive integers? Opportunities for student engagement now exist. Some groups use 1, 2, 3, and 4; other groups use 2, 3, 4, and 5; and others use 6, 7, 8, and 9. Are there regular things happening with each set of digits? Explain your findings.

At a higher level, algebra offers an abstract view of the problem. Let a represent the first digit. Then the remaining digits are $a + 1$, $a + 2$, and $a + 3$. The regularity found among the sets of four digits provides an opportunity for proof.

In our previous solution, 6 is the greatest digit and 5 is the next greatest. Using variables, we can represent the greatest two 2-digit number as $10(a + 3) + (a + 2)$. The format for the second 2-digit number is $10a + (a + 1)$. Subtracting, we obtain the following:

$$10(a + 3) + (a + 2) - [10a + (a + 1)] = 10a + 30 + a + 2 - 10a - a - 1 = 31$$

The result provides a solid argument that a difference of 31 can always be obtained with four consecutive digits with the format described.

Furthermore, teachers, mathematics leaders, and school leaders need a solid grasp of what each practice looks like when students are engaged in using the practice. Again, identifying actions that indicate the practice is being used in the classroom begins at one level, and it then deepens over time as both students and teachers gain proficiency.

Ultimately, change at a building or campus level resides within the domain of the principal. There are many outside forces at work that greatly influence change, but the principal has the duty, responsibility, and authority. This is a double-edged sword because, ultimately, the principal is, by position, responsible for everything that happens in the school, not just change initiatives.

Since these responsibilities are huge, principals need to distribute the responsibilities by becoming leaders of leaders. Often, this requires principals to encourage and support individuals in assuming leadership roles over specific areas or initiatives. The adoption of, and subsequent implementation of, the Common Core State Standards practically assures that successful principals need to use teaching staff to take charge of this mathematics change initiative. Doing so may not be how principals operated in the past, but there are too many necessary responsibilities and actions for one leader to handle.

Three possible committee arrangements are suggested—steering committee, leadership team, and learning communities. These committees are identified in generalized terms since campus and school sizes vary tremendously. Also, districts have various ways they provide personnel and assign their duties. The committees need to organize according to the resources principals have available. What is important is that leaders find arrangements that work for their unique situation.

Role of a Steering Committee

One of the first beneficial steps principals can make is to form a steering committee. The committee, made up of the principal, central office representative, mathematics teacher representatives, and hopefully a mathematics leader, helps organize and manage the data. The committee obtains and studies information related to adoption and implementation of the Common Core.

As the information is collected and studied, the committee identifies patterns and issues. These patterns and issues formulate into a general plan that highlights the significant changes that need to occur. Depending on the size of the school, the steering committee may continue to function, but another committee is also formed. This is the leadership team.

Role of the Leadership Team

The leadership team, made up of the principal, mathematics leader representative, and mathematics teacher representatives, develops a specific plan to address the needs identified by the steering committee. The committees need to share information and work collaboratively. Most likely, there will be some overlap in membership of the committees. After agreement has been reached and information shared, the principal may or may not temporarily dissolve the steering committee. The principal may also form another steering committee to address other emerging issues.

In our case, the leadership team specifically takes on responsibilities for implementing elements of the Common Core content and Practices. These responsibilities are related to the information and tools contained in this book. While leadership is shared, the principals are still responsible for the direction of the leadership team.

Role of the Principal

Principals set the tone, expectations, and conditions for the committees. Principals also control the resources that may be required. And principals are the grounding point for the vision and mission of the school. The committees do not in any way replace principals, but rather intentionally extend their reach. Principals do the following:

- Provide relevant and necessary information
- Focus on student achievement and success
- Insist that students be the beneficiaries of all actions
- Ensure committee recommendations are within the control of the school
- Ensure committee recommendations are operational within the school system

> ▶ Guarantee committees have the time and resources necessary
> ▶ Assist in moving action steps forward
> ▶ Help monitor implementation and progress
> ▶ Capture learning from successes and failures
> ▶ Sustain momentum

Principals provide leadership and promote leadership. Committee members may or may not have extensive backgrounds in working on committees. Principals must carefully nurture and mold team interactions for the members to be productive.

Developing Learning Communities

Learning communities may already exist within the schools. If so, then the steering committee and the leadership team provide focus and direction to the learning communities. Two-way communication flow is critical. If learning communities are not currently operating within the schools, principals, with the help of leadership teams, will need to begin making plans toward this end.

Forming learning communities is not the end of the work, but the beginning. Within learning communities, leaders must also exhibit leadership skills. Learning communities translate recommendations from the leadership team into classroom actions. Leadership teams and learning communities function to open classroom doors and remove teachers from working in isolation.

Working collaboratively within the classroom and in committees improves the culture of the school. Schools need to exude a culture of success that is communicated throughout the schools' zone of influence. Students develop and grow a culture of success when they work collaboratively in groups on meaningful mathematics. Teachers promote a culture of success when they work on various committees within the school and in learning communities. Leaders demonstrate a culture of success by bringing individuals into the decision-making loop. This sense of collegiality, responsibility, and success for all permeates the school and is readily noticed by visitors.

A Principal's Story

Bob B. is the principal of a K–5 elementary school with an average attendance of 500 students. Bob began his career as an elementary teacher, and after eight years in the classroom, he became the principal of the same school. He has been principal for 10 years. Bob is considered to be an excellent principal. He has the respect of his teachers, the superintendent, parents, and other principals.

Bob has focused his career on allowing his teachers to teach. He believes that his teachers know what to do for students, and his job is to preserve class time, control discipline, and work to resolve parental issues. While staying true to his beliefs, Bob is also very aware that mathematics success for the students in his school has remained elusive. Data from various sources, including state assessments, show that students' scores are gradually falling as test expectations increase.

Now, Bob's state has adopted the Common Core State Standards, and every district principals' meeting focuses on the forthcoming changes. Bob is particularly concerned when conversations revolve around the Standards for Mathematical Practice. Even though Bob does not routinely visit classrooms beyond required evaluations, he is attentive to the fact that the teachers teach and the students basically behave. Bob feels the pressure and knows "the clock is ticking." (Scenario continued in Chapter 4.)

Having Productive Conversations

1. Not everyone can serve on every committee, so how do you decide?
2. What is the balance between directions from the principal concerning the role of the committee and the committee's flexibility to decide?
3. How do the various committees or teams get formed, and in what order?
4. How do various committees or teams ensure everyone has a voice?
5. How do the various committees or teams actually take action?
6. How do the various committees or teams actually monitor actions taking place?
7. How do the various committees or teams receive feedback and increase success?

4

Rigor and the Standards for Practices

If the Standards for Mathematical Practice are going to be the leverage point for instituting mathematical rigor, then to actually implement the Standards for Practice in classrooms, teachers and leaders will need to carefully study them. The key is to extract guidance from the wording provided in the Common Core for each domain and the accompanying clusters where actions are specified. Once a basic grasp of the ideas is attained, teachers and leaders must delve deeper into their ideas and beliefs about each practice. Digging deeper into understanding the Practices evolves over time.

This chapter provides ideas about where to start in both defining the Practices and in recognizing the Practices in action. Also emerging is a realization that the definitions and actions are the same ones teachers undertake to ensure rigor. As indicated previously, the Practices are multilayered, and they unfold as deeper understanding is obtained through using them. Since it is not possible to offer every connection for every grade or course, we provide a starting place for teachers and leaders to initiate conversations. These collaborative conversations, as indicated by the chapter organization, must include what the Practices mean and what the Practices look like in classrooms.

Standards for Mathematical Practice

The Common Core writing group (National Governors Association Center for Best Practices and Council of Chief State School Officers, 2010) developed eight mathematical practices. According to the writers,

> The Standards for Mathematical Practice describe varieties of expertise that mathematics educators at all levels should seek to develop in their students. (p. 6; © Copyright 2010. National Governors Association Center for Best Practices and Council of Chief State School Officers. All rights reserved.)

These Practices guide how teachers should teach and students should learn mathematics. To better assist translating the Practices into specific actions, and to promote clarity, we have separated Practices 1 and 3 into parts "a" and "b." We did this so important information and knowledge would not be overlooked.

Practice 1a—Make Sense of Problems

Students are not able to accurately solve problems if they are unable to make sense of the problems. This does not mean to imply that students do not attempt to solve problems. Their attempts, while sometimes effective in getting answers, are frequently based on visual cues they extract from the problem. For instance, suppose students have been working on two-digit by one-digit multiplication. For practice, they are assigned word problems. Each word problem contains exactly one 2-digit number and one 1-digit number. Students scan the problem, select the numbers, and multiply. No reading, reasoning, or problem solving is required. Word problems of this nature do not promote sense making.

Defining the Practice

To demonstrate making sense of problems, students should do the following:

- Be willing to take time to carefully read and reread the problem, or listen as the problem is read to them.
- Be able to restate the problem using their words.
- Be able to recognize what information is provided and what additional information is needed.
- Be able to recreate the problem through drawings or manipulatives.
- Be able to identify important information and remove extraneous information.
- Be able to translate the problem into mathematical symbols.
- Be able to perform required operations and answer the problem's question(s).
- Be able to recognize and explain how they know an answer is reasonable and correct.

Recognizing the Practice in Action

To promote students' abilities to make sense of problems, teachers must provide numerous opportunities for students to work on problems that require thought. Students need experience with a diverse array of problems presented in varying formats. They need to work on solving the problems in both individual and collaborative settings. To make sense of problems, students will need time to share their strategies and solutions.

As teachers move about the classroom, they should observe and hear students discussing a problem. What are the key pieces of information? What is required? (Make sense of problems and persevere in solving them.) What operations are necessary? Can we solve the problem using manipulatives? (Use appropriate tools strategically.) Is an equation needed? (Reason abstractly and quantitatively.) Is our estimate reasonable? (Model with mathematics.) Answers to questions such as these provide good evidence that students are making sense of problems.

Practice 1b—Persevere in Solving Them

Several factors impact students' willingness to stay with a problem to successfully find a solution or several solutions. First, the problems must be challenging. Students do not learn to persevere by working one-step, simplistic word problems. Perseverance is far more likely to occur when students work in pairs or small groups. Perseverance is more likely when problems are relevant to students.

Defining the Practice

To demonstrate perseverance in solving problems, students should do the following:

- ▶ Be willing to read the problem more than one time.
- ▶ Be able to select a solution path and work through the problem.
- ▶ Be willing and able to modify or reject a solution path that is not working.
- ▶ Be able to analyze mistakes, correct the mistakes, and seek to solve the problem again.

Recognizing the Practice in Action

A variety of problem styles need to be used during instruction. Every problem should not be multiple-choice. Some problems should not provide answer choices. Other problems should be open ended so students are supplying the numbers and the answers. For instance, given a budget and a menu, students should offer several ways they could eat within the budget. Students could plan field trips or a class party.

Consider the following example where perseverance may be needed to find all solutions.

> Using the digits 0 through 9, create two 2-digit numbers that have a sum of 100. A digit can only be used once in a particular number sentence. Find all of the solutions.

$$\underline{\quad}\,\underline{\quad} + \underline{\quad}\,\underline{\quad} = 100$$

Students may select $21 + 79 = 100$ as a solution but then quickly realize that $31 + 69 = 100$ is also a solution. In fact, there are many possible solutions. Students discover that the digit 5 can never be used in the ones place since two 5s would be needed. Similarly, they find that the digit 0 cannot be used in any place. They begin to sort various solutions and arrange them possibly by place value. Students also find that the digit 9 cannot be used in the tens place since this would force the second number to be a single digit or would result in the number sentence $90 + 10 = 100$, where two zeros are used.

Practice 2—Reason Abstractly and Quantitatively

There are several components to mathematical reasoning. Mathematical reasoning is the ability to draw conclusions from given information or to recognize that insufficient information is provided for conclusions to be made. In the area of numbers, mathematical reasoning is using the properties of our number system to solve problems by translating a problem into symbols, following established rules to perform required calculations, and then translating the symbols back into a meaningful solution. Mathematical reasoning is further used when identifying the needed calculations. Finally, mathematical reasoning is used to affirm the answer is viable for the given problem. For geometry, mathematical reasoning involves drawing or constructing two- or three-dimensional figures with manipulatives, translating ideas or discoveries into symbols or equations, and solving problems quantitatively.

Defining the Practice

To demonstrate reasoning abstractly and quantitatively, students should do the following:

▶ Be able to separate multistep problems into smaller but meaningful parts.
▶ Be able to solve less complex versions of the problem, and use the obtained information to solve the existing problem.

- Be able to use drawings, manipulatives, or symbols to represent the problem.
- Be able to use symbols to represent the problem, and by looking at the symbols, be able to reconstruct the problem.
- Be able to recognize the derived solution is reasonable to the problem.

Recognizing the Practice in Action

Students must be provided multiple opportunities to share what they think concerning mathematical concepts, vocabulary, and operations. Their thinking must be valued, and their willingness to share their thinking appreciated. Once their thinking is identified, teachers must carefully guide the students by examples and counterexamples to refine students' thinking. As students explore and discuss examples and counterexamples, they increase their ability to reason. For example, student pairs may be asked to identify the similarities and differences between 2^2 and 2×2. Students would need to think about the symbols and what they actually mean. In geometry, consider Euler's Formula for solid figures that relates vertices, faces, and edges: $V + F - 2 = E$. Given one or two of the variables, can students reason abstractly to determine a figure or figures that can meet the conditions?

Practice 3—Construct Viable Arguments and Critique the Reasoning of Others

We believe that this particular practice should be discussed, and implemented, in two parts: 3a and 3b.

Practice 3a—Construct Viable Arguments

To construct viable arguments, students must be thinking and reasoning. Their reasoning is based on mathematical properties from many areas of mathematics. Constructing viable arguments is not developing a formal proof, but rather rationally explaining why a particular solution path was adopted, and why the path led to a reasonable answer. Making conjectures about observed patterns of some type and then constructing a logical explanation is a formidable, but necessary, task for understanding mathematics.

Defining the Practice

To demonstrate constructing viable arguments, students should do the following:

- ▶ Select a solution path and explain their reasoning for deciding on the path.
- ▶ Listen to other solution paths, and recognize if they are correct.
- ▶ Work collaboratively to reach consensus on a solution path by listening carefully to explanations and accurately explain their reasoning.

Recognizing the Practice in Action

Students must be discussing mathematics and mathematical problem solving. Students may engage in viable arguments with other classmates, teachers, or as leaders for the entire class of students. Students are explaining how they, or their group, arrived at the solution to a challenging problem.

An interesting problem in geometry for conjecturing and creating a viable argument involves the degree of a vertex. It is an area of three-dimensional geometry not found in textbooks. If the degree of a vertex is defined to be the number of line segments at a particular vertex and S represents the sum of the degrees, explain or create an equation that provides S for any prism. Students will observe by constructing models of prisms or drawing two-dimensional representations of prisms that every vertex has a degree of 3. For a cube or rectangular prism, $S = 8 \times 3 = 24$; for a triangular prism, $S = 6 \times 3 = 18$; for a pentagonal prism, $S = 10 \times 3 = 30$. Some students might provide an argument that to find S you double the number of sides of the prism base and multiply by 3, while others might argue that you multiply 6 times the number of sides of the base. Both arguments are correct.

Practice 3b—Critique the Reasoning of Others

Critiquing the reasoning of others is based on (1) students in the classroom are engaged in practices that require them to think and reason, and (2) students are sharing and reflecting on one another's explanations of their thinking and reasoning.

Defining the Practice

To demonstrate critiquing the reasoning of others, students should do the following:

▶ Show how two correct, but different, solution paths are similar and different.

▶ Be able to explain how a problem was solved after listening to the explanation.

▶ Given a disagreement on the solution, be able to defend why they believe their solution is correct and the other solution is incorrect.

▶ Listening to an explanation, identify accurate statements and inaccurate statements.

Recognizing the Practice in Action

When students are constructing and sharing viable arguments, they must then carefully consider the accuracy and reasonableness of solutions that differ from their own. When critiquing another's reasoning, a deeper understanding of the mathematics is required. Critiquing is not just stating the solution is correct or incorrect, but actually analyzing the entire solution and identifying what is correct or incorrect. If students are not sharing their thinking, they cannot be critiquing the reasoning of others.

Our previous geometry problem provides an excellent example where students might critique the reasoning of others. For one group, $S = 2 \times N \times 3$, where N is the number of sides of the base. For a second group, $S = 6 \times N$. Without much difficulty, it is easy to see that the reasoning of each group produces the same result for S.

Practice 4—Model With Mathematics

Students, when confronted with a problem situation, can transfer the mathematical knowledge they have into a solution path for the situation.

Defining the Practice

To demonstrate modeling with mathematics, students should do the following:

▶ Be able to represent a stated problem situation as mathematical symbols.

> ‣ Be able to recognize events in life as represented and solved mathematically.
> ‣ Be able to recognize, create, and use various data arrangements such as tables, graphs, and charts.

Recognizing the Practice in Action

Students need to be provided multiple opportunities to move between data arrangements and problem solving. When students are given a mathematical problem, they need to identify given data, generate unknown data, arrange the data, and solve the problem. Students must also be able to take data arrangements and configure reasonable situations that match the data.

Classroom experiences include opportunities for students to find solutions to problems that are of interest and grade/age appropriate. Students could plan a real or hypothetical party, trip, ceremony, or activity. Younger students could plan for the classroom, and older students could plan for the school or community. In the elementary grades, students begin to see relationships between two variables, an independent variable and a dependent variable, noting how events in everyday life capitalize on the relationships.

Students might create a "blueprint" for a classroom or building, using scaling techniques to denote lengths and widths.

Practice 5—Use Appropriate Tools Strategically

Students need to work with a variety of mathematical tools over extended time. For example, a one-week measurement unit is insufficient for providing students the knowledge they need to select and use tools, such as rulers, strategically.

Defining the Practice

To demonstrate using appropriate tools strategically, students should do the following:

> ‣ Be able to identify and describe various mathematical tools that are grade/age appropriate.
> ‣ Be able to accurately use a mathematical tool that is grade/age appropriate.
> ‣ Be able to select and use an appropriate mathematical tool for a specific problem situation.

▶ Be able to explain why or why not a particular tool is appropriate to the problem situation.

Recognizing the Practice in Action

Students must have ready access to a variety of mathematical tools such as rulers, compasses, models, protractors, counters, linking cubes, and calculators. Technology programs that demonstrate geometric or algebraic concepts are also invaluable. Beyond having these tools available, lessons must be planned and presented that ensure the tools are routinely used. When students are engaged in challenging problems, they should be able to have access to, experiment with, and use various tools in their efforts to find viable solutions.

There are an abundance of measurement examples where students use tools inappropriately. In observing students as they measure length with rulers or meter sticks, how do they align the tool with the segment? In measuring angles with a protractor, do they align the vertex with the "center" of the protractor? Do they measure an acute angle, but give the supplement because of the scale on the protractor?

Students using graphing calculators to graph functions sometimes fail to consider a "window" for a particular function. They observe a graph on the screen and accept it as correct without considering various constraints that are critical.

As we have noted, extended time is needed for students to use tools appropriately and strategically.

Practice 6—Attend to Precision

Students must recognize the value of being accurate and precise in their written language, discussions, and solutions. Students learn to appreciate mathematics as a symbolic language where the symbols have specific meanings and are used in standard ways to express and explain relationships as true.

Defining the Practice

To demonstrate attending to precision, students should do the following:

▶ Be clear and precise in their language and work.
▶ Be able to clearly and logically organize and deliver mathematical explanations.

▷ Be able to name and accurately use mathematical symbols.

▷ Regularly label units of measure correctly.

▷ Be able to clearly identify a problem solution, and justify the reasonableness of the solution.

Recognizing the Practice in Action

Students learn to attend to precision by regularly talking, listening, and restating. Teachers regularly model this practice by using precision in their language and assisting students in clarifying their explanations by restating. Teachers do not answer for the students, nor do they ignore incomplete student explanations. They probe with follow-up questions. For students to understand and articulate their thinking and reasoning, they must have multiple opportunities to think and reason. They must have opportunities to show labels, such as those involving measurement or graphing of data. Even with a simple number theory proof that the sum of two odd numbers is an even number, students should be able to show with manipulatives or logically explain that the statement is correct.

Practice 7—Look for and Make Use of Structure

Mathematics has structure, whether involving numbers or involving geometry. Within various number systems, mathematical operations are bounded. As students advance in their mathematical knowledge, the systems expand. Student conceptual understanding must also expand. Students perform operations with natural numbers, whole numbers, rational numbers, integers, irrational numbers, and complex numbers. Some properties of number systems work, and some do not. A simple illustration is the additive identity property. Since zero is not an element of the natural numbers, there is no additive identity property for this set of numbers. Geometry, especially transformational geometry, also provides a great deal of structure that relates to the structures found for number systems. With each advance, students need multiple opportunities to compare and contrast the structure of the new system to the previously learned one.

Defining the Practice

To demonstrate looking for and making use of structure, students should do the following:

- Be able to observe patterns in mathematics.
- Be able to understand when patterns are identified and explained.
- Be able to use mathematical structure to simplify and solve problems.
- Be able to conjecture that a pattern may exist, and test the pattern for reliability.

Recognizing the Practice in Action

Students need to demonstrate ease and fluency in using patterns to compute mentally. Simple counting patterns provide opportunities to see structure at a variety of grade levels:

- 5, 10, 15, 20, . . .
- 1, 2, 4, 7, 11, 16, . . .
- 1, 4, 9, 16, . . .
- 1, 5, 11, 19, . . .

In early grades, students know and use information such as 11 is 10 plus 1 more, and 12 is 10 plus 2 more. In later grades, students may observe the slope of a line and know both the steepness and direction. For geometry, they see that the number of lines of symmetry for a regular polygon is the same as the number of sides. By looking at patterns, they observe that that number of diagonals for a regular polygon can be determined using the formula $D = n(n - 3)/2$, where n is the number of sides of the polygon.

Practice 8—Look for and Express Regularity in Repeated Reasoning

Beginning in the elementary grades, students observe many different types of patterns in mathematics. With regularity in numbers, students adapt shortcut methods for calculations. In geometry, regularity in angle measurement provides a variety of opportunities for students to express ideas.

Defining the Practice

To look for and express regularity in repeated reasoning, students should do the following:

- Be able to conjecture and explain that a calculation repeats in some regular manner.
- Be able to understand various mathematical representations for indicating a repeating or growing pattern.

Recognizing the Practices in Action

When students are looking for and expressing regularity in repeated reasoning, they begin to understand the mathematical concepts they are studying. Some of the regularity starts in kindergarten and Grade 1, where students notice that even numbers and odd numbers alternate as they count on or back. We also noted the regularity in addition of even and odd numbers, which continues later with regularity in multiplying two odd numbers, two even numbers, or one of each. The square of numbers ending in five also provides regularity for computation. For example, $25^2 = 625$, $35^2 = 1225$, or $85^2 = 7225$. In each case, there is a 25. The number 6 is 2×3; the number 12 is 3×4; the number 72 is 8×9.

As students in Grade 5 begin dividing a unit fraction by a whole number, they notice a simple regularity when dividing by 2: $1/2 \div 2 = 1/4$, $1/4 \div 2 = 1/8$, $1/8 \div 2 = 1/16$, or even $1/3 \div 2 = 1/6$. The denominator of the quotient is the double of the denominator of the original unit fraction.

With fractions and decimals, students observe that if denominators have any factors other than two and/or five, the decimal repeats. Once again, they look for the regularity in the repeating decimals, such as $1/7 = 0.142857\ldots$; $2/7 = 0.285714\ldots$; $3/7 = 0.428571$; and so on.

In Euclidean geometry, students learn that the sum of the three interior angles of a triangle is always 180^0. Using this basic information, they see that a quadrilateral can be decomposed into two triangles, so the sum of the interior angles is 360^0; a pentagon can be decomposed into three triangles, so the sum of the interior angles is 540^0.

Rigor and Practices

After reading and reflecting on the Practices, notice that they reflect and contain a depth of learning. The Practices are not just surface-level topics that are easily learned and implemented, even though the surface level is, indeed, where the changes may actually start. Leaders and teachers note that students must be engaged in classroom discourse; however, they are unable to directly teach the Practices. The Practices continue to develop with every positive shift in classroom instructional strategies.

Rigor is thinking, reasoning, and depth of knowledge. The Practices strongly promote thinking, reasoning, and depth of knowledge; in other words, they promote rigor. Without the Practices, content is delivered without understanding. Consequently, leaders and teachers must consciously develop rigor by attending to implementation of the Practices over time and to depth.

A Principal's Story (Continued)

As principal, Bob has the opportunity to select and attend one professional development opportunity every year. He asked for, and received, permission to attend a National Elementary Principals Association meeting. The meeting brochure caught his attention because the focus of the program sessions was on meeting the challenges of implementing the Common Core. He knew he needed ideas.

At the meeting, he pointedly sought out sessions concerning the Standards for Mathematical Practice. Bob found the sessions informative. One particular session, however, was of particular interest. The session discussed the meaning of the Practices, the actions students must be taking in mathematics classrooms, and provided a Proficiency Matrix with a recommended instructional strategies sequence. The Matrix was used to guide professional conversations about instructional change. Bob decided that he and some of his key teachers would study the Proficiency Matrix and find ways to use it to support change. (Scenario continued in Chapter 5.)

Having Productive Conversations

1. Are some Practices more important to you?
2. In your opinion, which Practices must students master to be proficient in mathematics?
3. How might a Practice look different at various grade levels or subject areas?
4. What actions are students generally taking in the classroom if the Practices are evident?
5. How frequently do various Practices occur in classrooms? Daily? Weekly? Monthly?

5

Rigor Related to Classroom Formative Assessment

In Chapter 1, *Understanding and Meeting the Challenge of Rigor,* our nation is described as one driven by testing, and school systems are a very real part of the phenomenon. Students are constantly tested in schools. The tests include daily quizzes, weekly tests, chapter tests, unit tests, six- or nine-week tests, semester tests, final exams, and now new end-of-course exams. Beyond these classroom requirements, schools also participate in state-level tests, and perhaps district-developed benchmark tests. Test giving and test taking are normal routines within schools. Significant instructional time is allocated for students taking tests as well as time and resources for grading tests. Every type of test has some form of data generated.

If generating data from tests were the solution to students' mathematics learning difficulties, we would surely be at the top of international comparisons of students' test results. We are not at the top, and in fact, we are not near the top. According to the National Research Council STEM Report (2011),

> For example, as measured by the National Assessment of Educational Progress, roughly 75 percent of U.S. 8th graders are not proficient in mathematics when they complete 8th grade. Moreover, there are significant gaps in achievement between student population groups: the black/white, Hispanic/white, and high-poverty/low-poverty gaps are often close to 1 standard deviation in size. A gap of this size means that the average student in the underserved groups of black, Hispanic, or low-income students performs roughly at the 20th percentile rather than the 50th percentile. U.S. students also lag behind the highest performing nations on international assessments: for example, only 10 percent of U.S. 8th graders met the Trends in International Mathematics and Science Study advanced international benchmark in science, compared with 32 percent in Singapore and 25 percent in China. (p. 3)

Although the recent TIMSS report showed improvement in mathematics for Grade 4 and Grade 8 students, with both averages above the international averages, our overall rankings still did not place us near the top (Provasnik et al., 2012, pp. iii–iv).

With all of this data being gathered about student performance, how is such a discrepancy explained? Does it not seem reasonable for leaders and teachers to precisely know students' individual strengths and weaknesses? Wouldn't students also know? Prepared with this information, remediation should be a snap. Why does this abundance of data not result in student mathematical success? Perhaps we need to better understand assessment, instruction, data, and feedback to unravel this dilemma.

Assessment Types

We deliberately use the term "test" to describe the current actions taking place in schools. Testing and test results serve a multitude of purposes, some beneficial, and some perhaps not beneficial. There is now a shift in terms to emphasize differences. We use the word "assessment" to indicate when the data, regardless of how the data are derived, are used to inform students of their learning and teachers about instruction. Furthermore, the term "assessment" is intended to indicate more rigor—more depth of understanding. We are testing students, but are we assessing them? In other words, do students and teachers receive adequate and timely feedback that informs both about the students' current understanding of the mathematics, and is the feedback being used for this purpose?

To initiate instructional change, teachers need to understand various assessment types. Understanding assessment builds the foundation for enacting manageable, effective change. There are, essentially, three forms of assessment: formal, informal, and formative. Formal and informal assessments constitute *summative* assessment measures.

Summative assessments, consisting of both formal and informal assessment, are designed to check for student learning **after** instruction has occurred. This "after" heavily relies on when and how the feedback is returned to students. For discussion purposes, we define *formal* and *informal* assessments as follows:

> *Formal assessments:* Valid, reliable tests with analysis and data feedback on results. These usually include state-developed and administered tests. Other tests include CAT, SAT, ACT, or other standardized achievement tests. Formal tests have items that have been carefully screened and selected.

> *Informal assessments:* Scored or graded tests. These may have group data analysis. The items have not been validated or checked for reliability. These tests usually include district benchmark assessments, six- or nine-week tests, semester exams, or teacher-made tests.

Both of these summative assessment types have feedback returned to students within a time span of days to months.

Opposed to summative assessments are those assessments that occur immediately **during** classroom instruction. This difference is critically important. We define *formative* assessment as follows:

> *Formative assessment/ongoing formative assessment:* Data collected and used within the instructional classroom. These assessments draw from observations, questions, and work analyzed during instruction.

Feedback for formative assessments should be given before students leave the classroom or shift to other subject areas. They need to receive feedback in a timely fashion to make use of it. In addition, teachers are not the only source for formative assessments. Students play a significant role through collaboration and self-monitoring.

In staying with this idea, according to Heritage (2011, p. 18), feedback has two aspects:

1. Feedback obtained from planned or spontaneous evidence is an essential resource for teachers to shape new learning through adjustments in their instruction.
2. Feedback that the teacher provides to students is also an essential resource so the students can take active steps to advance their learning.

The term "spontaneous" is critical. Spontaneous evidence is indicated by Brookhart (2008) who poses a question about when to give feedback and then responds:

> When would a student want to hear feedback? When he or she is still thinking about the work, of course.

Feedback is also better understood and incorporated when students are actually thinking about the mathematics.

Classroom Formative Assessment: The Missing Instructional Element

As a rule, formative assessment (also referred to as ongoing formative assessment) is randomly defined by educators. Ideas and opinions abound. We have purposefully defined formative assessment as assessment occurring *during* classroom instruction, and have also relegated formal and informal assessments to the summative category. This decision to refine formative assessment is not without research-based support, nor is it intended to imply that summative feedback is unimportant. The distinction is made for a reason. If, as we imply in our definition of rigor, that it is the effective confluence of content and instruction, then formative assessment is an invaluable link between the two.

Refining Formative Assessment

There is an urgent need to truly believe that all students can learn to think mathematically. This position—that all students can learn to think mathematically—is supported by the National Research Council (NRC, 2004) and the National Council of Teachers of Mathematics (NCTM, 2000). The National Research Council also maintains that students' thinking is vitally important to learning. In *How Students Learn* (NRC, 2005) they state:

> Students' thinking resides at the instructional center; therefore, teachers must regularly take stock of it and make it visible. (p. 192)

To this end, students' thinking must be made overt and visible (NRC, 2000) since feedback, based on their thinking, must be provided. Frequent feedback allows students to observe their progress and, more important, self-correct misunderstandings (NRC, 2004). Both assessments and feedback need to focus on students' understanding of the learning (NRC, 2000). "Although formal paper-and-pencil assessments are useful strategies for gauging student learning, they provide limited information" (NRC, 2004, p. 84).

For feedback to be beneficial, the National Research Council believes,

> The roles of assessment must be expanded beyond the traditional concept of testing. The use of frequent formative assessment helps make students' thinking visible to themselves, their peers, and their teacher. (2000, p. 19)

The National Mathematics Advisory Panel (2008) echoes this belief when the panel states, "Formative assessment—the ongoing monitoring of student learning to inform instruction—is generally considered a hallmark of effective instruction in any discipline" (p. 46).

These beliefs weave together to form a strong position for supporting students' thinking and reasoning as a central theme to mathematical learning. Students' thinking and reasoning are fundamental precepts for mathematical rigor. Further, the students' thinking and reasoning must be overt and visible **during** instruction so appropriate, timely feedback can be provided. Because monitoring through testing is already occurring in abundance, then, we hold forth that assessment of students' thinking during instruction—classroom formative assessment—is the missing ingredient to learning success as well as the launch for mathematical rigor.

As a result of our conclusion and classification, tests, and even assessments, are numerous, yet classroom formative assessments, according to our definition, are rare. The effective use of ongoing classroom formative assessment is the *missing element* to effective instructional change. For whatever reason, formal and informal assessments have failed to provide significant change in how students are taught mathematics. While formal and informal assessments have most likely impacted the mathematics content taught—the "what," they have not created an observed need to change "how" mathematics is taught. When mathematical rigor is related to the Practices, the "how" mathematics is taught strongly influences the "what" mathematics is taught.

Classroom Formative Assessment

Far too often, formal and informal assessments are administered as multiple-choice tests, or as ones where only the answer is "graded." According to Boaler (2008), this narrow testing format dramatically limits the content assessed as well as the thinking required to be successful (if indeed thinking is even required). She states,

> In addition to a reliance on multiple-choice formats, the mathematics tests used in most states across America are extremely narrow. They do not assess thinking, reasoning, or problem solving. (p. 87)

Thinking, reasoning, and problem solving are indicators of mathematical rigor. However, we often remove any problems that actually require students to think and reason in the mistaken belief that more challenging problems would only frustrate the students. Our problems focus only on rules and procedures. Without challenging problems and deep thinking, rigor is absent.

On the other hand, classroom formative assessments are rarely multiple-choice. Formative assessments are a variety of techniques used to currently determine students' thinking and understanding of the content being taught. The distinguishing characteristic of formative assessment within this book is that the collected information is used immediately to either affirm correct understanding or to intervene and correct misunderstandings. This distinction is vitally important. When confronted with student misunderstandings within the classroom, teachers are compelled to shift their instruction, the "how."

Since classroom formative assessments are used as part of the lesson, most students do not leave the classroom with confusion about mathematical content. Also, since teachers are constantly checking on class understanding, students are more highly engaged and not allowed to "sit out" of the lesson. By holding students more accountable and having them highly engaged, students are more closely attending to the content and the instruction.

In conjunction with the development of the Common Core content and Practices, assessments are being rethought and reconfigured. Assessments are being designed to determine students' abilities to think and reason while solving challenging problems. These forthcoming assessments are certainly more rigorous. The trend reinforces the need for teachers to promote students' visible thinking through classroom formative assessment. In all likelihood, future assessments will easily match and track student performance on rigorous mathematics content, including thinking and reasoning through the lens of the Practices.

Formative Assessment and Intervention

As stated earlier, classroom formative assessment is a key factor to effective instructional change, thus, improving student learning and supporting rigor. Why? The focus is now on student actions first, and then teacher reactions. Formative assessment must include intervention or feedback to have the desired results. When formative assessment focuses on gathering information about student understanding, teachers must actually use the information to transform their instruction in addressing issues as they arise.

Meeting the demands for increased student learning and coping with the increasing pressure related to student performance, teachers need to focus on two dimensions of formative assessment—understanding current student learning and providing appropriate feedback and intervention.

Current Learning

The term "visible thinking" is used to convey the idea of understanding the current level of student learning. As the term implies, students' thinking must be obvious and overt. Students must be required to think and learn to express their thinking in ways that make sense to themselves, their classmates, and their teachers. This level of thinking serves as a self-monitoring mechanism. It also provides teachers opportunities to either affirm students' understanding or to immediately intervene to correct students' misunderstanding.

Visible thinking, formative assessment, and intervention are not the dominant themes for most mathematics classrooms. Using the most commonly adopted lesson format, teachers follow a pattern of brief explanation, procedure model, procedure-guided practice, and independent practice. Questions focus on procedures and are directed at one student at a time. This lesson arrangement is so prevalent that the National Research Council (2001) named it "recitation." In *Adding It Up*, the NRC states:

> What is most striking in these observers' reports is that the core of teaching—the way in which the teacher and students interact about the subject being taught—has changed very little over that time. The commonest form of teaching in U.S. schools has been called recitation. Recitation means that the teacher leads the class of students through the lesson material by asking questions that can be answered with brief responses, often one word. (p. 48)

This approach to teaching and learning will not work to promote the Practices and will, therefore, fail to promote mathematical rigor.

Effective Intervention

Like attaining rigor, learning to use effective intervention is a process. It takes focus, energy, and time. The results, however, are outstanding. Intervention works in direct collaboration with visible thinking and formative assessment to increase classroom rigor. Much like formative assessment, intervention may occur in three separate venues: outside the classroom; inside the classroom, but outside the regular lesson; or inside the classroom during the lesson. We are only discussing intervention within the regular classroom during routine instruction.

Intervention, however, must be better defined for our purpose in this book. Intervention is often considered as "reteach." For reteach, students mostly failed to perform correctly and the teacher essentially teaches the same lesson again. In some cases, replacement lessons are actually used, but the instructional approaches are the same. Intervention may also be viewed as additional practice or work outside the regular classroom. This is not the meaning of intervention we are using for this book. For our purposes, intervention occurs within the classroom. Teachers affirm correct understanding and ensure all students are aware of the affirmation, or they correct misunderstandings by making instructional adjustments.

Intervention, correcting or affirming understanding, may occur in numerous ways. First, teachers are not always the dispensers of the correcting or affirming. Second, correcting or affirming should not be predictable by the students. For instance, teachers should routinely ask students to explain their thinking. This request for additional explanation should occur equally for cases where students are correct and incorrect. If not, students quickly learn that additional thinking is only required when the answer is incorrect.

Asking for additional explanation is an intervention. Allowing students to work with a partner to share and discuss their individual answers is an intervention. Teachers may ask students to demonstrate their answer using models or pictures as a way to intervene. Teachers may have several students come to the board to share different strategies and answers as an intervention. A variety of questions designed to elicit thinking can be posed:

- ▶ Can you tell me more?
- ▶ Did you see a pattern? If so, what did you see?
- ▶ Does that always work?
- ▶ Can you think of a different way?
- ▶ You and (student's name) worked the problem in different ways, but your answers are the same. How can that happen?
- ▶ Can you explain how (student's name) worked the problem?
- ▶ That's a good question. What do you think?
- ▶ Talk it over with your partner. Do you agree on your answers?

Teachers may use an example problem that has been solved incorrectly as a guide for classroom discussions. Teachers state that a student from last year worked the problem in a particular way, and the answers were incorrect. An example is provided in Box 5.1.

Box 5.1

Teacher: Students, look at the problem from last year's class, and the answers a student provided.

The problem: Find the answers: $2^3 = ?$ $2^4 = ?$

Student answers: Since $2^2 = 4$, then $2^3 = 6$ and $2^4 = 8$.

What do you think the student misunderstood? If you could, how would you explain this problem to the students to help them understand?

Talk with your partner and develop a strategy with drawings or pictures to help this student understand.

Finally, classroom formative assessment changes "how" teachers teach. Strategies are used that promote student engagement, thinking, and reasoning. Strategies are also used to make the students' thinking and reasoning overt or "visible."

Visible thinking may be described as clarity and transparency in one's cognitive processes (Hull, Balka, & Harbin Miles, 2012, p. 2). For students learning mathematics, this clarity and transparency is a result of teachers constantly monitoring student understanding (formative assessment) and providing immediate feedback (immediate intervention) during the instructional time. There is, of course, a caveat. For teachers to actually monitor student thinking and reasoning, students must be provided opportunities to think and reason.

Ongoing formative assessment and intervention are well worth the required instructional time. When teachers discover that a student does not understand or lacks some critical element of understanding, there is a high probability that other students also have this same lack of clarity. A brief pause to correct misunderstandings or affirm understandings can have a true ripple effect.

Furthermore, as teachers experience using these strategies, and clarifying understandings, they more intentionally plan lessons that promote using the strategies and building understandings. Teachers are the key to student success.

Instructional Research

Research has long acknowledged the vital role of the classroom teacher related to student learning (Reeves, 2006). Primarily, the research actions teachers take are directly aligned to what occurs within the classroom instructional period. Further review of research strongly indicates that successful instruction is predicated on students being actively engaged in the learning activities designed and carried out by teachers. Indicators of active engagement include the following:

- Students are talking and sharing (NRC, 2004).
- Students are collaborating in small groups (NRC, 2004).
- Students are thinking and reasoning (NRC, 1999).
- Students are explaining their thinking (NRC, 2000).
- Students are using manipulatives to construct meaning (NRC, 2000).
- Students are devoting appropriate time to solving challenging problems (NRC, 2000).

When studying or reviewing effective research-supported instructional strategies, the obvious becomes clear. Each strategy targets and promotes a student's ability to think and reason. Thinking and reasoning are at the heart of the CCS Standards for Mathematics Practice. They are also the actions required for mathematical rigor. Further, the effective research-informed practices and strategies actively engage students in instructional activities. Why is this so? They have students engaged, reasoning, thinking, and making sense of mathematical content.

Teachers should recall strategies such as think, pair-share, higher-order questioning, wait time, collaborative grouping, sharing thought processes, and justifying reasoning. Perhaps teachers are familiar with the NCTM process standards: problem solving, reasoning and proof, communications, connections, and representation. Maybe they review the eight Standards for Mathematical Practice from the CCSS. In every case, students are the focal point. Students are called on to think and reason, be aware of their thinking and reasoning, and be aware of their classmates' thinking and reasoning. Students are also called on to make their thinking and reasoning visible to their teachers, to their classmates, and to themselves.

These are the same types of actions required by the Standards for Practice and the same required for attaining mathematical rigor. Furthermore, these research-supported actions are the same as those required by classroom formative assessment, visible thinking, and intervention. Understanding the power of classroom formative assessment and how it links to the Practices, the Matrix, and rigor is essential to implementation.

The research strategies may be exemplified by a problem that requires teacher and student interaction as offered in Box 5.2.

Box 5.2

Domain: Number and Operations—Fractions

Standard: ***Use equivalent fractions as a strategy to add and subtract fractions***. Solve word problems involving addition and subtraction of fractions referring to the same whole, including cases of unlike denominators (e.g., by using visual fraction models or equations to represent the problem). Use benchmark fractions and number sense of fractions to estimate mentally and assess the reasonableness of answers.

Problem

Mary and Sara want to go to summer camp. A local company has offered to hire potential campers. The company will pay for picking up trash along local

(Continued)

(Continued)

roadways. The company will pay $24.00 per mile in quarter-mile segments for both sides of the road.

Part 1. To afford to go to camp, Mary and Sara must each earn $105.00. On how much roadway do Mary and Sara have to pick up litter?

Part 2. Sara picks up litter on 3/8 of a mile, and Mary picks up litter on 5/6 of a mile. How much distance did they cover together? If Sara picks up litter on 9/16 of a mile and Mary picks up litter on 7/8 of a mile. How much distance did they cover?

Part 3. Sara and Mary each take one side of the roadway. After working for the first mile, Mary discovers that she is faster than Sara. At the end of the mile, Mary is 1/4 mile ahead of Sara. If Sara picks up litter at the rate of 1/4 in 20 minutes, how fast does Mary pick up litter?

Part 4. At the end of 5 miles, Mary moves to Sara's side of the roadway and picks up litter until the girls meet. Approximate how they should fairly share the money? Demonstrate your reasoning and explain your decision.

(Note: students are not expected to solve the problem algebraically, but rather by estimation and approximation.)

Part 5. Mary and Sara are working on three different sections of roadway. They have completed one section of 2 1/2 miles, and a second section of 4 1/8 miles. How long is the last section of roadway if Mary and Sara both earn the money they need for camp (there are several considerations and arrangements possible)? Determine the length of roadway you determined, and explain your arrangement and reasoning.

Part 6. If Mary and Sara worked apart, how long did each one spend picking up litter?

If Mary and Sara worked at their own pace, but Mary helped Sara, how long did they pick up trash?

If Mary and Sara worked together at Sara's pace, how long did they pick up trash?

Part 7. What does Mary do after finishing her section? Do you think Mary helps Sara? What would you do? Why? What is fair? Why do you think your answer is fair?

Part 8. Mary and Sara found a sponsor that would pay an additional $10.00 for every mile of litter they picked up. If Mary and Sara worked together at Sara's pace, how many quarter-mile sections of roadway would they need to pick up litter?

As students work on engaging problems, such as Figure 5.2, the effective strategies and Practices emerge. For instance, students are challenged to make sense of the problem and to persevere in solving the various sections. Students must reason, model, and attend to precision while engaged in the problem.

Synergy

The whole is greater than the sum of its parts. While the parts may be discussed or explained individually, they work as a whole. The Proficiency Matrix is introduced and described in Chapter 6. It is designed to impact classroom instruction directly related to teaching the CCSS mathematics content through the Standards for Mathematical Practice. For the Matrix to be used as designed, three important and interrelated concepts have been addressed in this chapter:

- ▶ Classroom ongoing formative assessment
- ▶ Visible thinking
- ▶ Intervention

As noted, these concepts are the critical elements currently missing or inadequately addressed in many mathematics lessons. When missing, rigor is missing; when present, mathematical rigor is present. The Practices will not be attained without the concepts being understood and implemented. As students are engaged in activities that require thinking and reasoning, teachers are engaged in activities that make students' thinking visible. In this way, teachers are able to quickly affirm appropriate understanding, as well as immediately correct misunderstandings.

By instituting classroom formative assessment, teachers are far more aware of students' understanding, their thinking, and their reasoning—the skills called for in the Practices. This awareness of students' understanding by teachers demands that teachers take action immediately.

Classroom ongoing formative assessment is the cornerstone for a successful change initiative that meets the demands of the Standards for Mathematical Practice. Incorporating classroom formative assessment is realistic and achievable. Teachers, mathematics leaders, and school leaders need to focus their energy and resources to making classroom formative assessment a reality in every mathematics classroom. To affirm an important point, ongoing formative assessment is both "what" (content) and "how" (instruction) teachers teach. In this context, ongoing formative assessment is not a pen-and-paper test administered to students.

A Principal's Story (Continued)

When Bob returned from his national meeting, he immediately made copies of the Matrix for several of his key teacher leaders. He knew his mathematics teachers already had copies of the Standards for Practice, but he believed the Matrix truly helped explain what the Practice statements mean. He then contacted the mathematics specialist assigned to his campus to set up a one-on-one meeting. Before the meeting, he asked the mathematics specialist to think about the Practices and how the specialist believed the Practices would look like when implemented.

He told the specialist he was sending her a document through the school mail, and he sent her a copy of the Proficiency Matrix.

At the meeting with the mathematics specialist, Bob shared his concerns about the gradual decline of mathematics scores. Even though his school was doing OK, he knew serious problems were coming in the future if instruction did not change. He also asked about progress at other campuses. He told the mathematics specialist, "I have to do something, and I need your help. I would like for you to help gather information you think we will need, and I am going to approach several of my teachers for help."

Over the next week, Bob privately talked with four teachers he felt were respected by his other teachers. He explained his concerns and pointed out that change was coming in one way or another. Bob believed, and he told his teachers, that it would be best to get ahead of the change. He also stated that their conversations were not to be kept secret. He needed suggestions and he trusted his teachers to provide them.

Finally, Bob asked each teacher if they would commit to studying the Practices. If they were expected to implement them, it would be best to truly understand them. (Scenario continued in Chapter 6.)

Having Productive Conversations

1. What assessment types are currently in use by your school or district?
2. How are data from these assessments reported to students?
3. How do teachers use data from these assessments?
4. How do leaders use data from these assessments?
5. Which data source(s) can most accurately portray student learning?
6. What data can formative assessments provide teachers?
7. What assessment combinations could best inform instructional practices related to student learning?

6

Rigor and the Proficiency Matrix

For readers unfamiliar with the Proficiency Matrix, it is a document designed to promote attainment of the Standards for Mathematical Practice and, therefore, classroom formative assessment. As students and teachers progress through the levels of proficiency, mathematical rigor is instituted and attained. The document is intended to promote collaborative, productive conversations between teachers and leaders that focus on student learning and success in mathematics. In this chapter, we explain the organization of the Matrix, and prepare teachers and leaders to effectively use the Matrix.

Organization

The Proficiency Matrix (Table 6.1 and Appendix A) is a rubric arranged in rows and columns for ease of use. The form may be printed as a two-sided document. In the left column are the Standards for Mathematical Practice numbers with Practices 1 and 3 separated into two parts, "a" and "b." The second column is the practice wording from the Common Core. The next three columns signify levels or degrees of proficiency for individual practices. The levels or degrees progress from Initial to Intermediate to Advanced. For instance, Practice 1a: Make sense of problems, is shown below:

	Students:	(I) = Initial	(IN) = Intermediate	(A) = Advanced
1a	Make sense of problems.	Explain their thought processes in solving a problem one way. (Pair-Share)	Explain their thought processes in solving a problem and representing it in several ways. (Question/Wait Time)	Discuss, explain, and demonstrate solving a problem with multiple representations and in multiple ways. (Grouping/Engaging)

Each of the other practices follows the same format.

The statements in the cells serve as brief indicators and reminders of the practices, and they offer a progression of student depth in both understanding and engagement. These indicators, due to their brevity, must be placed within the context of a grade level or subject area.

Initial

Once teachers begin focusing on implementing the practices in their classrooms, they realize students must be required to talk about what they learned and how they understand the mathematics. This student talk is elevated beyond reciting a series of procedural steps used to solve a closed problem. Since teachers cannot listen to every student at the same time, students must talk with other students. Talking with another student about one's thinking requires listening to one's partner's thinking.

Intermediate

As students become comfortable with the idea of sharing their thinking and listening to another's thinking, the ability to compare and contrast solutions or solution strategies emerges. Students begin to see that "a" solution approach is not "the" solution approach. As a result, students are able to analyze various solution approaches. Students also learn to seek other solution routes, even after they have discovered one that works.

Advanced

Students at the advanced proficiency are able to carefully analyze their thinking, construct logical explanations, and justify their reasoning in clear, precise language. Students move fluidly between words and mathematical symbols. They are also able to comfortably explain the relationship among mathematical concepts. These student abilities, when combined with the appropriate Common Core mathematics content, produce a mathematically rigorous course.

Progress Toward Rigor

We stated earlier the fundamental premise that rigor requires significant shifts in instruction and, therefore, is an ongoing change process. As instructional techniques shift, classroom climate and culture shift, student achievement and performance expectations increase, and rigor emerges. Mathematically rigorous classrooms are the outcomes as students and teachers work through the proficiency levels and strategies identified within the Matrix. As students and teachers gain expertise with the Standards of Practice, and as they move toward the advanced level of implementation, mathematics classes increase in rigor. Just as there are no shortcuts to meaningfully implement the Practices, there are no shortcuts for reaching and maintaining rigor.

Strategy Relationship in the Matrix

The Proficiency Matrix (Table 6.1) includes one additional piece of information to assist teachers in meaningfully implementing the Practices and attaining rigor. In each cell of the Matrix beneath the levels or degrees is assigned an instructional strategy. Two very important items need to be explained. First, our intent is not to imply that one single strategy serves to accomplish the level of proficiency needed to master a practice. The suggested strategy is an exemplar of what works well. The strategy promotes the indicator and is correlated to the indicators on the Classroom Visit Tally—Teacher form discussed in Chapter 11. In addition, the strategy listing is sequential and developmental. After each strategy is learned, incorporated, and perfected, the next strategy naturally arises. The strategies, in sequential order, are as follows:

- Initiating think, pair-share;
- Showing thinking in classrooms;
- Questioning and wait time;
- Grouping and engaging problems;
- Using questions and prompts with groups;
- Allowing students to struggle; and
- Encouraging reasoning (Hull, Balka, & Harbin Miles, 2012)

A brief description for each strategy is located in Appendix B.

We described mathematical rigor as the confluence of content and instruction. The Matrix assists in integrating the Practices and instruction in a way that can be effectively used during lesson planning. By gradually incorporating these strategies over time, teachers directly impact student levels of attention and engagement. Furthermore, as each successive strategy is perfected, students attain higher degrees of proficiency. With each practice being attained at higher levels, student thinking, reasoning, and understanding of mathematics is achieved. With these advancements, students attain mathematical rigor.

Classroom Formative Assessment and the Matrix

Classroom formative assessment—checking student understanding continually during classroom instruction—requires active engagement by students. To assess students' understanding, thinking, and reasoning, students must be engaged in ways that visibly demonstrate their thinking, reasoning, and understanding. In essence, students must be discussing, sharing, collaborating, and attending to a lesson. The Standards for Mathematical Practice point the direction for what students do, or actions they take, while learning mathematics. The Matrix expands on and clarifies actions for the practices. When students are engaged in these actions, their thinking is made visible, and teachers can assess students' depth of comprehension. In other words, the indicators identified in the Matrix provide the actions students must be doing for teachers to engage in classroom formative assessment. As students and teachers undertake these actions, mathematics classrooms increase in rigor.

The Matrix helps guide lesson planning and student actions that need to occur during instruction. While the Matrix is not a template used to plan lessons or a checklist of actions to be taken, it does serve as a reflective tool. Optimally, the Matrix is used during collaborative lesson planning. The lesson that follows provides an example of how teachers use the Matrix to guide their thoughts and actions in preparing for how students will be engaged during the lesson.

Lesson Example: Using the Matrix to Select Strategies and Student Actions While Planning

Ms. Edwards's Classroom

Ms. Edwards teaches fifth grade. She has been working to incorporate the Standards for Practice as she teaches the Common Core mathematics content. She has discovered that students are appearing to really enjoy mathematics class more. The students are comfortable working in pairs, so Ms. Edwards wants to try a more engaging problem with students working in groups of three. She wants her students to think and reason more deeply, model with mathematics, and critique the reasoning of the other students. Looking at the Proficiency Matrix, she focuses on the following:

Practice 2. Reason abstractly and quantitatively for initial and intermediate levels.

Initial: Reason with models or pictorial representations to solve problems.

Table 6.1 Proficiency Matrix (also in Appendix A)

	Students:	(I) = Initial	(IN) = Intermediate	(A) = Advanced
1a	Make sense of problems.	Explain their thought processes in solving a problem one way. *(Pair-Share)*	Explain their thought processes in solving a problem and representing it in several ways. *(Questioning/Wait Time)*	Discuss, explain, and demonstrate solving a problem with multiple representations and in multiple ways. *(Grouping/Engaging)*
1b	Persevere in solving them.	Stay with a challenging problem for more than one attempt. *(Questioning/Wait Time)*	Try several approaches in finding a solution, and only seek hints if stuck. *(Grouping/Engaging)*	Struggle with various attempts over time, and learn from previous solution attempts. *(Allowing Struggle)*
2	Reason abstractly and quantitatively.	Reason with models or pictorial representations to solve problems. *(Grouping/Engaging)*	Translate situations into symbols for solving problems. *(Grouping/Engaging)*	Convert situations into symbols to appropriately solve problems as well as convert symbols into meaningful situations. *(Encouraging Reasoning)*
3a	Construct viable arguments.	Explain their thinking for the solution they found. *(Showing Thinking)*	Explain their own thinking and thinking of others with accurate vocabulary. *(Questioning/Wait Time)*	Justify and explain, with accurate language and vocabulary, why their solution is correct. *(Grouping/Engaging)*
3b	Critique the reasoning of others.	Understand and discuss other ideas and approaches. *(Pair-Share)*	Explain other students' solutions and identify strengths and weaknesses of the solutions. *(Questioning/Wait Time)*	Compare and contrast various solution strategies, and explain the reasoning of others. *(Grouping/Engaging)*

(Continued)

Table 6.1 (Continued)

	Students:	(I) = Initial	(IN) = Intermediate	(A) = Advanced
4	Model with mathematics.	Use models to represent and solve a problem, and translate the solution into mathematical symbols. *(Grouping/Engaging)*	Use models and symbols to represent and solve a problem, and accurately explain the solution representation. *(Question/Prompt)*	Use a variety of models, symbolic representations, and technology tools to demonstrate a solution to a problem. *(Showing Thinking)*
5	Use appropriate tools strategically.	Use the appropriate tool to find a solution. *(Grouping/Engaging)*	Select from a variety of tools the ones that can be used to solve a problem, and explain their reasoning for the selection. *(Grouping/Engaging)*	Combine various tools, including technology, explore, and solve a problem as well as justify their tool selection and problem solution. *(Allowing Struggle)*
6	Attend to precision.	Communicate their reasoning and solution to others. *(Showing Thinking)*	Incorporate appropriate vocabulary and symbols in communicating their reasoning and solution to others. *(Allowing Struggle)*	Use appropriate symbols, vocabulary, and labeling to effectively communicate and exchange ideas. *(Encouraging Reasoning)*
7	Look for and make use of structure.	Look for structure within mathematics to help them solve problems efficiently (such as 2 × 7 × 5 has the same value as 2 × 5 × 7, so instead of multiplying 14 × 5, which is [2 × 7] × 5, the student can mentally calculate 10 × 7.) *(Question/Prompt)*	Compose and decompose number situations and relationships through observed patterns in order to simplify solutions. *(Allowing Struggle)*	See complex and complicated mathematical expressions as component parts. *(Encouraging Reasoning)*
8	Look for and express regularity in repeated reasoning.	Look for obvious patterns, and use if/then reasoning strategies for obvious patterns. *(Grouping/Engaging)*	Find and explain subtle patterns. *(Allowing Struggle)*	Discover deep, underlying relationships (uncover a model or equation that unifies the various aspects of a problem such as discovering an underlying function). *(Encouraging Reasoning)*

Source: © LCM 2011, Hull, Balka, and Harbin Miles, mathleadership.com

Intermediate: Translate situations into symbols for solving problems.

Practice 3b. Critique the reasoning of others for intermediate level.

Intermediate: Explain other students' solutions and identify strengths and weaknesses of solutions.

Practice 4. Model with mathematics for intermediate level.

Intermediate: Use models and symbols to represent and solve a problem, and accurately explain the solution representation.

She is planning for the lesson to take two class days but is okay with extending into three if class discussion about the mathematics is rich.

She has taken a more traditional problem about two boys mowing lawns and extended the problem. She plans to have students demonstrate their work on large sheets of paper, and then hang the work around the room. Students then take a gallery walk comparing different ways groups solved the problem. She will then consecutively number each group's paper.

After time for the groups to review the work, she is sending the students back to their small groups of three to offer their thoughts on the following:

A) Which group's work was accurate?
B) How can the different approaches (even if the solution was incorrect) be clustered, along with a written explanation of why the work was sorted as such?
C) If a group's solution was thought to be incorrect, where was the error made? The small groups were to select only one group's work to review if the solution was thought to be incorrect.

Finally, Ms. Edwards intends to pull several groups that used different approaches to the front of the classroom to explain their work and solution.

Ms. Edwards wants her students to understand that a solution approach could be correct even though a calculation was incorrect. She wants to emphasize the importance of effort and thinking. Furthermore, Ms. Edwards plans on carefully monitoring group work during the exploration phase. While mistakes could slip through, she intends to intervene with groups who are working the problem incorrectly. She recognizes this is a balancing act of providing guidance and probing questions without directly providing a singular solution path and answer.

Ms. Edwards plans to emphasize rigor by integrating content and instruction through using the Matrix.

Domain: Geometry

Standard: Graph points on the coordinate plane to solve real-world and mathematical problems. Represent real-world and mathematical problems by graphing points in the first quadrant of the coordinate plane, and interpret coordinate values of points in the context of the situation.

Problem

Two brothers, Jerry and John, work together to mow lawns. Because Jerry is the older brother, he can mow faster than John. To be fair, they share the money earned by the area of yards each brother mows.

For 1/4-acre lots, Jerry mows 5/8 and John mows 3/8. They are paid $44 for mowing 1/4-acre lots.

Part 1. Groups are to construct a graph or graphs demonstrating the relationship between the boys' mowing and their earnings.

Part 2. Using the graph, the groups are to determine what is the approximate amount of money each boy earns for mowing half-acre lots? How do you know? What are the pros and cons of using a graph for this information? Can you determine a better way to display the data? Why is your way better?

Part 3. It takes about 1 hour 15 minutes for the boys to mow a half-acre lot. What can you determine with this additional information?

Part 4. Are you able to determine how much each boy earns per hour? If so, determine the hourly rate, and create a graph to demonstrate.

Part 5. Are you able to determine how long it takes each boy to mow a 1/4-acre lot? If so, determine the answer and create a graph to show their time for mowing lots of various sizes.

Part 6. Both Jerry and John are saving for his own bicycle. If they can spend a maximum of $219 each, without tax, how many hours must each boy mow? How many acres does each boy need to mow?

The problem scenario that we have described illustrates a major change in mathematics instruction. In many classrooms, only the original sentence with numbers would be used. Students would be required to determine how Jerry and John shared the $44 (5/8 x 44, 3/8 x 44). No graphing of information would be involved. Now, students are involved with graphing points in the plane and looking at a variety of possible ratios.

A Principal's Story (Continued)

Bob scheduled several meetings with his selected teachers and his mathematics specialist. He held a few meetings after school, but he also arranged for meetings during the school day. At the first meeting, everyone agreed to take two or three Practices to collect information on and to study. With six people, every Practice had at least two people collecting information on them.

At later meetings, the Practices were discussed, with the individuals responsible for studying them leading the conversations. For each Practice, the group began making lists that would help define what the Practices would look like when being used inside classrooms. With each successive meeting, clarity about the Practices was gained, and details were emerging concerning actions students and teachers should be taking.

At the end of the discussion over the Practices, the leadership team related the list and discussions to the Proficiency Matrix. Plans emerged for how the team could use the information they had obtained as well as use the Matrix. In no uncertain terms, the leadership team recognized that students must become more actively engaged during classroom instruction. A plan was developed to increase student engagement through working in pairs. Team members also put together their thoughts about seeking or creating more engaging materials that provided students the opportunity to think and reason. The change process was under way. (End of scenario.)

Having Productive Conversations

1. What do you like about the Matrix?
2. What do you dislike about the Matrix?
3. What could you do to enhance the Matrix for your use?
4. In the teaching cycle (plan, present, analyze, and reflect), how can the Matrix be used to support learning?

PART II

Issues and Obstacles

As with any significant program calling for substantive change, leaders and teachers face numerous issues and obstacles. This is just as true for implementation of the Common Core. Without care being taken, these issues and obstacles can seriously interfere with moving adoption and implementation forward. Issues that arise may divert necessary time and focus from the actions required to successfully launch or sustain program changes. Obstacles may promote a feeling of despair and hopelessness since efforts are thwarted, and a feeling of "what's the use" can emerge.

Issues in this section are related to the Common Core and are important topics that need to be addressed during implementation. Some preceded the Common Core initiative. Obstacles arise with the program or are systemic habits that block implementation. In some situations, it may become difficult to cleanly separate issues and obstacles since issues may become obstacles. We discuss some of the most important issues including understanding the Practices and meeting the demands of diversity. The obstacles of working in isolation, attempting to evaluate people to change, and failing to monitor student actions are discussed in the chapters in a sequenced manner so leaders and teachers can successfully manage them.

7

Issues to Resolve

Several issues reside with any change in a mathematics program or instructional approach. The issues presented here are related to one another. They evolve around student success, progress, and achievement in mathematics. The three issues must be carefully considered and managed during a change process. The issues also directly relate to criteria used to identify a rigorous mathematics program. The issues are the following:

▶ Teaching the identified content
▶ Deepening mathematical understandings
▶ Reaching all students

Issues are not dealt with one time and resolved. Even though the issues may be resolved, they must be continually monitored to ensure they remain resolved.

Issue: Teaching the Identified Content

One of the most important issues related to mathematics achievement is in guaranteeing that the mathematics content intended to be taught is the content taught (English, 2000; Marzano, 2003). This includes more than just covering topics or domains, but teaching for the depth indicated by the written curriculum and the CCSS.

Each grade level or subject for mathematics, starting in kindergarten and ending in exit-level high school courses, has an abundance of content. As a result, teachers have little time to spend on "nonproductive" activities. Clearly, teachers do not have time to teach last year's content before teaching this year's content. Students must be provided appropriate opportunities to learn the content for their grade level or subject.

Students need to begin mathematics instruction on day 1, and the content must be the correct content. Obviously, review of some content and skills is required, but this review must be scaffolded within the instruction rather than reviewed over several weeks at the beginning of school. Beginning-year reviews of this nature waste time. If students do not know or remember the required content, they will not know it at the appropriate time either since there is a gap of time between being reminded and needing the skill or knowledge. Further, reviews of this nature reinforce learning content and skills as isolated segments, rather than as a body of connected knowledge.

One of the greatest difficulties mathematics teachers and leaders face is ensuring every student is taught the identified content. Content disparity is frequently the basis for achievement gaps between subgroup populations.

Issue: Deepening Mathematical Understandings

Deepening mathematical understandings primarily focuses on students and what must occur for students to meaningfully learn mathematics, build proficiency with the Practices, and acquire appropriate and useful mathematical knowledge. While the focus is on students, it is obvious that teachers must also expand their understanding of mathematics. In particular, teachers must better relate pedagogy and content. They must also deepen their understandings of how students learn mathematics concepts, where students frequently develop confusion, and how most efficiently to correct the confusions. Finally, teachers need to continually increase their knowledge and awareness of the developmental nature of mathematics.

Making Connections

Mathematics teachers and leaders must help students make connections between the daily objectives, weekly objectives, unit objectives, and even the overall course or subject area objectives. The content taught, by necessity, is presented in a linear manner. Hopefully, the content is arranged in a sequence that is actually developmental. Nonetheless, just because adults have designed a sequence that makes sense to them does not mean students intuitively understand the sequence. Teachers must be overt about how the mathematics unfolds.

Making connections across time and content involves making meaning. Connections among mathematics concepts need to be explicit. Connections are more than skills increasing or reminding students of previous procedures and definitions. Students need to purposefully discuss how each instructional unit relates to past learning and how the unit prepares the students for future learning.

Meeting the needs of diverse learners means teachers purposefully connect learning opportunities.

Creating Meaning

Students must understand mathematics as a subject that makes sense, has purpose, and is meaningful. Students who struggle with mathematics rarely remember the rules or understand the litany of procedures. They form the overwhelming belief that mathematics is some mysterious code that only a few gifted students can decipher. According to Boaler (2008, p. 43),

> Children begin school as natural problem solvers and many studies have shown that students are better at solving problems before they attend math classes. They think and reason their way through problems, using methods in creative ways, but after a few hundred hours of passive math learning students have their problem-solving abilities drained out of them. They think that they need to remember the hundreds of rules they have practiced and they abandon their common sense in order to follow rules.

Once students begin struggling, this belief is reinforced by our current approach for intervention and remediation. Rather than show students the purpose of mathematics, we insist they practice more rules and procedures in preparation for learning meaningful mathematics. Since the rules and procedures are already confusing to these students, they fail to improve in performing the procedures. As a result, they are not introduced to meaningful mathematics, and the self-fulfilling loop is complete.

This scenario is the exact opposite of what should occur. How can students make sense of mathematics when all they see are disconnected rules and procedures that, to them, make no sense?

Meeting the needs of diverse learners requires that teachers ensure their lessons demonstrate mathematics as meaningful and useful, and they must include realistic, challenging problems.

Using Learning Research

Attention spans for children and adults are limited. While older students do have the ability to attend for longer than do younger students, the length of time, on average, is about 8 to 10 minutes. If students can only focus their attention for 8 to 10 minutes, what happens after 10 minutes? Basically, students tend to "zone out." Even though they may be quiet, they are not attentive, and certainly not understanding.

Students who have greater difficulty with mathematics content or who have been convinced they cannot learn mathematics lose interest more quickly. For this reason, teachers must plan lessons that include frequent mental breaks that allow students to process what they have heard or experienced. Any break in the routine is helpful, but allowing students time to talk and share is far more successful.

There is an exception to the attention span rule, or at least a variation. Students working together on interesting, engaging problems can continue their work past the 10-minute time. This variation is, naturally, related to the age of the students, previous opportunities provided to collaborate, and the task required.

If classrooms are going to meet the needs of diverse learners, then the instructional strategies must actively engage students. Gavin and Moylan (2012) provide several recommendations. First, they recommend using appropriate tasks, ones that stress concept development and push students to work beyond their comfort zone (p. 185).

Another suggestion strongly recommends that students be required to discuss concepts and justify their reasoning. They encourage teachers to have students communicate through writing, where writing includes drawing representations (p. 186). This recommendation is backed up by their request that teachers use formative assessment to inform instruction (p. 189). The Proficiency Matrix and the Strategy Sequence support these recommendations.

Issue: Reaching All Students

There is growing diversity within U.S. schools and mathematics classrooms. Population trends clearly indicate more diversity lays in the future, not less. Teachers need to know how the Matrix and the Practices support teachers in meeting the needs of all students rather than complicating the process.

Learning Opportunities

Meeting the needs of diverse learners means students are provided every opportunity to learn the appropriate mathematics content to the appropriate depth of understanding. Since students learn at different rates and under different conditions, teachers and leaders must ensure students have the required time to learn the mathematics. Beyond time, teachers must be supplied with the appropriate tools.

Several suggestions for meeting the needs of all students have already been broached. Obviously, implementing the Standards for Mathematical Practice as described assists in meeting students' learning needs. Classroom formative assessments that make student thinking visible provide teachers plentiful opportunities to meet students' needs. Also, rigor, approached as depth of thinking, reasoning, and understanding, promotes success for every student. According to the Literacy and Numeracy Secretariat (2008, p. 1),

> The purpose of differentiating instruction in all subject areas is to engage students in instruction and learning in the classroom. All students need sufficient time and a variety of problem-solving contexts to use concepts, procedures and strategies and to develop and consolidate their understanding. When teachers are aware of their students' prior knowledge and experiences, they can consider the different ways that students learn without predefining their capacity for learning.

Slicing to the core of this statement brings about the understanding that students must be engaged, students must be required to think and reason, and teachers must know what the students understand.

Indicators in the Matrix, and the suggested strategy for each indicator, greatly assist teachers in meeting the needs of English Language Learners (ELL) and other underserved students. These same indicators and strategies assist teachers in differentiating classroom instruction for students operating at various mathematical learning levels. Consider this simpler version of a grid puzzle from *Visible Thinking Activities* by Balka and Hull (2011).

The 3 × 3 grid here contains 9 numbers that are related in various ways according to the following characteristics:

Prime numbers	Even numbers
Multiples of 3	Numbers of the form $3n + 1$
Numbers greater than 9	Square numbers

2	3	4
7	9	10
12	16	18

Rearrange the numbers in the blank grid and write the characteristics on the left and at the top so that the number in a cell satisfies both conditions.

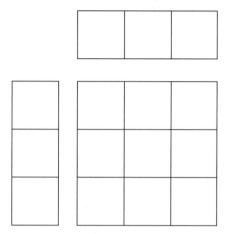

The task involves a great deal of mathematical content and vocabulary related to number sense. Most students randomly write numbers in the cells without any initial categorization of the numbers. Using a pair-share strategy, one student might choose to find all the prime numbers in the original grid, while another finds all the numbers greater than 9. They discuss and explain their thought processes for placing numbers in the categories and continue sharing with the remaining categories. At the same time, they are looking for obvious patterns. Some students will be stuck at this point and not complete the task. They are at the Initial level in the Matrix. Others will begin looking at the intersections of the various categories. They are attending to precision by using appropriate vocabulary. They observe that certain categories only contain three numbers and that a common element appears in these pairs of categories.

Prime numbers	2, 3, 7
Even numbers	2, 4, 10, 12, 16, 18
Multiples of 3	3, 9, 12, 18
Numbers of the form $3n + 1$	4, 7, 10
Numbers greater than 9	10, 12, 16, 18
Square numbers	4, 9, 16

The students have made use of structure but still have not completely solved the problem. This group of students is at the Intermediate level of the Matrix. Finally, there are groups of students who are at the Advanced level of the Matrix. They explain that if there is an overlap in two categories that have only three numbers, then one characteristic must be on the horizontal and the other on the vertical. Students begin filling in the grid until it is complete.

	Prime	Square	> 9
3n + 1	7	4	10
Even	2	16	18
Multi	3	9	12

They justify and explain why their solution is correct, communicating various paths they took to complete the grid puzzle. This problem, although seemingly simple, involves many of the Practices and provides the engagement called for in CCSS.

As teachers observe students engaged at different levels in this particular task, they will be able to differentiate instruction to meet varying student needs: What are the multiples of 3? What is a square number? What are the prime numbers? Do you notice anything about certain groups of numbers? What does it mean if two characteristics share the same number? How will you represent that result on the grid?

In a great number of schools, far too many students are relegated to below-grade-level courses in a fruitless hope that slowing down will somehow result in speeding up. These students are routinely taught and retaught basic computational skills. This approach constantly widens the performance gap. The Common Core content fueled by the Practices offers a very real opportunity to finally overcome this issue to student success in mathematics.

Using the Strategy Sequence to Address Issues

Understanding and using the Strategy Sequence Chart (Appendix B) is very important for implementation of the Standards for Mathematical Practice. Two questions are considered:

- ▶ How is it a sequence?
- ▶ Why is it a sequence?

The answers to these questions directly relate to effective mathematics instruction and meeting the needs of diverse learners.

How is the chart a sequence?

Teachers and leaders need to read and discuss the Strategy Sequence Chart provided in Appendix B. Seven categories are identified:

Instructional Strategy Sequence of Inclusion

- Initiating think, pair-share
- Showing thinking in classrooms
- Questioning and wait time
- Grouping and engaging problems
- Using questions and prompts with groups
- Allowing students to struggle
- Encouraging reasoning

These seven categories blend strategies and teacher actions. They imply student actions. In addition, there is not a clear, definitive line between each category. There is, however, a definite purpose in the design. The categories unfold as teachers gain experience with the various strategies associated with each category. The chart has a specific launch point—initiating think, pair-share.

To change behaviors, people must start where they are, then incrementally change and perfect skills over time. Rarely, if ever, is a skill used perfectly the first time. This statement is as true for teaching as it is for other careers. To increase student engagement, teachers must allow students to talk and to share ideas and understandings with their classmates.

The significant first step in the process and progression is allowing students to periodically talk to a student partner. The duration, at the beginning, may be as short as 15 seconds, but the step is critical. If teachers are unable to let students talk together for a short period, they are not going to successfully have students working in collaborative groups on challenging problems.

Looking at the Strategy Sequence, note the first three categories.

Instructional Strategy Sequence of Inclusion

- Pair-share
- Sharing thinking in classrooms
- Wait time and questioning

They all relate to students working with partners. As students begin discussing their ideas and relating their understanding, teachers grow in their ability to extract clearer explanations and delve deeper into students' thinking. This initial interaction eventually leads teachers to asking more challenging questions and providing opportunities for students to think before responding.

The next four categories require students to work in groups of three or four. These categories are identified by a significant shift in student responsibility for learning. Teachers routinely shift to facilitators of learning rather than dispensers of learning. This shift requires a difference in the type of work students are expected to complete. If students are to think and reason, they must have materials and problems about which they can think and reason.

- Grouping and engaging problems
- Using questions and prompts with groups
- Allowing students to struggle
- Encouraging reasoning

Why is the chart a sequence?

There are multiple reasons for the Strategy Sequence Chart to be a sequence. As already indicated, change is a process and progression for teachers. Also, change is a process and progression for students. Students who have not been asked, nor expected, to understand mathematics are not going to be able to instantly hold forth with valuable and thoughtful discourse on mathematical topics. Students are accustomed to explaining procedures they used to solve problems. They are not accustomed to explaining their thinking. In fact, they are probably unaware of their thought processes and the fact that they are even thinking. Students must develop this metacognitive realization. Consider mathematical problems that have multiple solutions. After finding a few solutions, students often start listing and making tables. They make conjectures about certain conditions that exist for the problem. If manipulatives are available, they model the mathematics for a solution and then unknowingly use their thought processes to find additional solutions. In other words, they make sense of problems and persevere in solving them, and they model with mathematics.

Embedded within the Strategy Sequence is a time factor that provides needed time for students and teachers to adjust to new instructional techniques and new learning demands. The sequence, when followed, actually works to change the learning culture. As teachers tend to work in isolation, so do students. Collaborating is new, and students must learn they can trust the other students in their classroom.

Finally, the chart is a sequence because it relates to the Proficiency Matrix. The Matrix pushes students to increase their thinking and reasoning while learning mathematics. The Matrix provides a developmental look at students as they grow in their ability to deal with more sophisticated concepts. This growth is achieved as teachers use instructional strategies that demand deeper mathematical understanding.

The Strategy Sequence is designed so each category is inclusive of the previous categories. In other words, teachers gain in the use of additional strategies rather than replace strategies. The sequence is intended to be a comprehensive "tool kit" of strategies that teachers can continue to add to over time.

Using the Matrix

Once familiar with the Matrix, teachers can routinely use information in the cells to analyze and reflect on their students' performances in obtaining the Practices. When meeting the needs of diverse learners, teachers want to pay particular attention to what proficiencies individual students are able to successfully demonstrate.

As students increase their ability to work with partners and small groups, teachers increase their ability to conduct classroom formative assessments by observing and listening. While circulating about the room, teachers notice if students are using models and tools appropriately. They hear what students are saying and which students are doing the talking. Teachers observe student work and quickly assess the degree of understanding demonstrated.

When students are sharing their solutions, teachers are able to ask specific questions targeted to various students to ensure understanding. Teachers have many opportunities to guarantee all students are learning.

As facility with the Matrix increases, teachers begin planning more explicit lessons designed to purposefully highlight particular proficiency cells. During this planning time, teachers are able to use the Matrix to identify specific strategies they wish to use to assess and promote understanding for identified students.

In closing, one critical point must be clearly understood. The Standards for Mathematical Practice require inclusion of different classroom instructional strategies than those currently being used. Failure to include additional strategies that require students' thinking to be made visible means the Practices will not be implemented and rigor will not be achieved—no exceptions.

Having Productive Conversations

1. How are you arranging to meet the needs of all students?
2. Does your data indicate that your ELL students are successful?
3. How does the Strategy Sequence address the need to differentiate?
4. How can the Matrix provide differentiated instruction?

8

Obstacles to Success

With any change initiative, there are obstacles that work to thwart the change. Most, if not all, of these obstacles are well known and well documented. Knowing, however, does not mean the obstacles are resolved. Leaders and leadership teams must directly work to address these obstacles for implementation to be successful.

Often, these obstacles reside within current system habits or operating procedures. As we have already described, systems seek balance. The actions of teachers, leaders, and students have been honed over time to produce the current mode of operating. Significant, important change requires a change in the daily mode of operation. Five significant obstacles are addressed in this chapter: isolation, evaluation rather than support, failure to monitor, overadoption, and mistaken efforts.

Obstacle: Working in Isolation

School personnel overwhelmingly tend to work in isolation when involved with their professional duties. Even though meetings may be held, and conversations occur, these activities are not directly related to the events happening within the classrooms.

Overcoming isolation means that ongoing events regularly occur within schools that allow teachers to collaboratively and intentionally work on mathematics lessons to be taught, analyze and reflect on lessons already taught, use student work to inform instructional practices, and open up their classrooms to visitors.

Isolation can be overcome with professional learning communities, mathematics specialists assigned to work directly with classroom teachers, and classroom visits for monitoring purposes. Even though all of these options may not be readily available to schools, variations must be created. If teachers are required to return to their individual classrooms and privately initiate change, then change will not occur.

Obstacle: Attempting to Evaluate People to Change

Evaluation is the general default setting for dealing with people in transition. Evaluations have been formalized, and educators are accustomed to the process. However, evaluation is not designed to change practices. Evaluation is a rating system that compares individual performance against a list of standard criteria. One only needs to meet the standard to be viewed as acceptable. Evaluation standards, once set, rarely change. Certainly, the standards do not change with each adopted initiative.

Formal teacher evaluations have been required for at least 40 or more years. While perhaps not categorically studied through research, it does seem that if evaluation were going to change instructional practices, the change would have already taken place. This is obviously not the case.

Evaluation models are not designed as classroom support for change. In evaluation, the evaluator has no contact with the teacher during instruction. Usually, the evaluator sits in the back of the classroom, observes the lesson, and takes notes. The notes are later transferred to some standardized formal form with a rating system. Teachers are presented a copy of the form during a post conference, and the score is explained.

Evaluation forms strongly reflect system philosophy. In an effort to fairly evaluate and compare teachers, forms frequently record only the actions a teacher took during the lesson—what the teacher said and what the teacher did. These actions rarely relate to student reactions to the strategies used. While there may be a few generic items on the form for student actions, they are perhaps related to discipline or demeanor. The message is clear—teachers are the center, not student learning. Obviously, one or two of these types of observations a year will not significantly impact instructional change, and likely hinder change initiatives. According to Murphy, Hallinger, and Heck (2013), in their research on the effectiveness of teacher evaluation, they find,

> Equally important, there is a robust body of empirical work that informs us that if school improvement is the goal, school leaders would be advised to spend their time and energy in areas other than teacher evaluation. (p. 352)

Evaluations are an obstacle to change initiatives. This is not a statement about the worth, value, or necessity of evaluations. The point is that formal evaluations are not agents of change. Change requires ongoing support and nonevaluative feedback. Any change initiative related to rigor must focus on student learning.

Obstacle: Failing to Monitor Student Actions

Failure to monitor implementation and progress practically ensures a lack of change and success. Program evaluation is one form of monitoring, but additional monitoring must occur. Monitoring processes collect data about student learning to inform teachers and leaders about the degree of change, the necessary support needed, and if the desired actions are indeed the acquired ones. Monitoring data that indicates the desired actions are occurring supports evidence of whether the actions are having a positive effect on student learning.

Obstacle: Overadoption

One obstacle that frequently occurs is unfortunately self-imposed. This obstacle arises when optimism and excitement overrule reasonableness. Certainly, optimism and excitement are desirable, but so are realistic expectations. Adopting and working to implement three significant changes and accomplishing none of them is not more desirable than adopting one change and having it successfully implemented.

Sometimes, leadership teams feel they must please a principal or an administration. Perhaps, they feel the need to move faster. In this pursuit, they decide to take on several significant changes at once. Change is difficult, and attempting multiple changes at once does not fit what is known about how adults change. This situation is particularly true at the initial stages of a significant change. Leaders are admonished to start slow and then build speed from success.

There is good news. When considering the Strategy Sequence, leaders and teachers can observe that different strategies unfold at different rates due to prior success in implementing a strategy. Also, if a strategy has been adopted and implemented, such as questioning and wait time, then expansions on this strategy are more easily assimilated.

Obstacle: Mistaken Efforts

Of all the obstacles, this is the perhaps the most disastrous. Far too often, the conclusion is reached before the exploration is even started. A change initiative has been identified in some way, and that is the initiative that is adopted. This process demeans leadership teams and reinforces the idea that they are not seriously valued. Solutions should not be offered or discussed before the real problems have been uncovered.

In other situations, mistaken efforts wander far afield from what will actually help students. There are no quick fixes when improving instruction. Changing class schedules, rearranging the order of courses, or changing the length of courses without regard to substance creates a sense of accomplishment, but no results. This process often leads to a sense of defeat and hopelessness. Worse, it can lead to blame. If all the effort has not produced benefits, then someone must be at fault—parents, students, teachers, or administrators.

Leadership teams and leaders need to focus their efforts on students. What are the events taking place that prevent many students from learning? These efforts are not targeted at blaming students and arriving at solutions, such as students need to do their homework. These are not "if only" wish lists either. Rather, the conversations focus on what the adults can do to increase student learning by addressing the research on how students best learn mathematics.

The obstacles will naturally arise as the change process is initiated. The obstacles discussed are frequently blocks to progress. Issues may sidetrack efforts by compelling leaders to address an issue in isolation or by initiating a subprogram that could likely interfere with implementation of the Practices. Leaders and teachers need to know how the Practices and the solutions address each of the issues. In the next chapters, solutions are provided with the appropriate tools to support implementation. When engaging in the solutions, leaders and teachers are addressing the issues and overcoming the obstacles.

As a note of interest, previous efforts at instituting rigor have most likely fallen under this category of "mistaken efforts." Rigor was identified as a missing element, and then misguided efforts added more content to mathematics courses or grade levels without any expectations to change instruction. In other cases, approaches intended to increase higher-order thinking were superimposed on the existing curriculum. Change was not recognized as a process, and the efforts were ineffective.

Mathematics Adoption Analysis Tool

As a way to assist leadership teams in identifying, adopting, and implementing worthwhile strategies, innovations, or programs, the Mathematics Adoption Analysis Tool (MAAT) is available. This tool, Figure 8.1, identifies

Figure 8.1 Mathematics Adoption Analysis Tool

Directions: Complete the form to the degree possible. Time may be appropriately spread over several days. The length of time required to complete the form is in direct relationship to the scope of the strategy, innovation, or program. Usually adopting a strategy requires the shortest amount of time while adopting a program may require the most time.

Strategy	Innovation	Program
Skill or technique teachers use within the classroom such as think, pair-share, questioning, wait time	Instructional materials requiring multiple strategies or new learning such as an interactive whiteboard or calculators	Significant shift in curriculum documents, materials, and resources, such as textbook adoptions or supplemental resources

Describe in concise terms. Be specific.

1. What is the strategy, innovation, or program under consideration?	
2. What are the critical components of the strategy, innovation, or program (i.e., concept-based, hands-on, manipulatives, independent work, technology-based, etc.)?	
3. What must teachers know and do to effectively use the strategy, innovation, or program?	

4. What must students know and do to effectively use the strategy, innovation, or program?							
5. Are there similar or competing strategies, innovations, or programs? If so, what?							
6. What resources are absolutely essential to effectively use the strategy, innovation, or program?							
7. What research-based best practice(s) can form the basis for the strategy, innovation, or program?							
8. When the strategy, innovation, or program is effectively implemented, what exactly will it look like in classrooms? What are teachers doing? What are students doing?							
9. What are the specific student benefits derived from using the strategy, innovation, or program?							
10. Do these specific benefits support the CCS Standards for Mathematical Practice? How? If the innovation or program provides content, does it align to the Common Core?							
11. How will student learning be monitored and assessed (directly related to student knowledge and student benefits)?							

(Continued)

Figure 8.1 (Continued)

Decision Point 1	Is this a worthwhile strategy, innovation, or program?
	Date _____ Approved by _____
	What training and support do teachers need? Others need?
	How will the strategy, innovation, or program be sustained?

	What is the minimum required commitment to the strategy, innovation, or program? **Time: initial and sustained** **Financial: initial and sustained** **Materials: initial and sustained**
Decision Point 2:	**What percentage of the individuals studying the strategy, innovations, or program actually support adoption?**
	Member **Yes** **No** **Undecided**

Recommendation _____   Final Approval _____

important criteria for leadership teams to consider when attempting to lead change efforts. MAAT is used when schoolwide implementation of an initiative is intended. It may be adapted according to the strategy, innovation, or program being considered. Certain strategies may be quickly discussed and approved without every question being answered. Innovations, in some cases, may also have selected questions deleted. Programs, most likely, require sufficient time and care, so all questions are considered. In every case, all strategies, innovations, and programs must directly benefit students.

Understanding MAAT

The Mathematics Adoption Analysis Tool is designed to assist leadership teams and leaders in avoiding the issues and obstacles that frequently undermine change efforts. The form also highlights the change process by identifying elements that indicate support and time needed for promoting and sustaining change. The form is intended to be completed in a collaborative process even though individuals may be responsible for certain sections.

After completing the sections of the form, a decision should be reached and an implementation plan outlined, the innovative ideas adjusted, or the ideas abandoned. If the ideas are adopted, then during the course of implementation, the form should be revisited and revised as new information emerges. The form maintains a focus on critical elements of the adopted change, and it serves as a monitoring tool to guide appropriate implementation. If, for example, the strategy, innovation, or program requires specific materials as resources for effective implementation, and those specified resources are not available to the teachers, the change fails. Likewise, if the materials are purchased, but not used, the change again fails.

Having Productive Conversations

1. Can you think of some potential obstacles you may encounter?
2. What do you believe is the best way to effectively handle obstacles?
3. What might you do to intervene in some possible obstacles before the obstacles create a larger problem?

PART III

Solutions

Successfully implementing the Standards for Mathematical Practice demands attention from school leaders, mathematics leaders, and teachers. There are five solution steps with each step being an iterative process. The steps are based on the Practices, the Proficiency Matrix, and the Strategy Sequence. These steps provide specific information concerning the current strategy usage within classrooms. The information drives professional learning decisions and informs the next steps in the improvement process.

9

Solution Step 1

Monitoring Student Actions
Related to the Practices

Once desired actions have been determined, leaders and teachers must monitor their efforts at change and clearly document progress and results. Classroom formative assessment and implementation of the Standards for Mathematical Practice will not occur by accident. Purposeful and strategic steps with specific actions must be undertaken. In addition, teachers and leaders must ensure that the current systemic operating conditions appropriately shift to accommodate new changes. The starting place for change is with the actions students take within the classroom.

Leaders, including teachers, monitor changes in student actions through a nonevaluative, supportive role. This method is contrary to the usual approach of evaluating teachers who are working to incorporate instructional change.

Failure to monitor change efforts assures complete failure. Lack of effective and supportive monitoring is perhaps the most severe barrier for leaders and teachers to overcome. Since the Practices focus on students' actions in classrooms, we initiate monitoring recommendations on these student actions. If monitoring is occurring, then classroom doors must be open.

Opening Classroom Doors

As the challenges to meet the demands of the Common Core are addressed, leadership teams are faced with a significant, yet often unnoticed, barrier. This barrier is a statement of fact that is perhaps unpleasant to accept. The fact is,

> *Systemic conditions are perfectly balanced to support the current leader, teacher, and student actions.*

School systems' operations have gradually evolved and been refined over decades. Systems effectively sustain current operating conditions. They are in harmony or stasis. To change classroom instructional strategies, systems must be shifted to support the desired change. This barrier's existence is often denied or ignored to the detriment of change. Perhaps the most significant systemic condition is the perceived notion of classroom sanctity. Individual classrooms are viewed as the sole domain of the classroom teacher, rather than the students being seen as the responsibility of the school. To enact effective change, this "closed door" policy must cease to exist. While this does mean that leaders have more access to classrooms, it more importantly means that teachers easily flow in and out of classrooms.

Nonevaluative Monitoring

For teachers to meaningfully engage in actions that support formative assessment, they must receive support and encouragement. They cannot be expected to make the required shifts by working in isolation from one another. Furthermore, since formative assessment occurs during instructional time, they must receive support during instructional time.

This point raises several key issues. Support is vastly different from evaluation. Consequently, formal school leaders responsible for evaluation are left with a dilemma. The question—how can leaders provide support for change without crossing into evaluation? The answer—they must build leadership teams with teachers.

An additional and highly effective method that most nearly replicates formative assessment is the use of mathematics coaches. By coteaching lessons, teachers and coaches, working in a collaborative and trusting relationship, are provided insight into formative assessment from the instructional side.

Nonetheless, coteaching is not always practical. Clearly, a coach cannot be assigned to every classroom every day. Yet teachers need data to inform their practice. This data can be obtained through classroom visits. Classroom visits, as used in this book, are nonevaluative, do not record teachers' names, and are conducted by various professionals.

Starting With Students

The focus of the Practices and the Matrix is on students and the actions they undertake during instruction. For this reason, classroom visits begin by focusing on student actions. After all, the students are the ones who need to be engaged in learning mathematics. Also, teachers plan and present their lessons. They already know what actions they are taking. What is far less clear is what the students are doing, and most critical, what the students are learning. Finally, our focus on improvement has spotlighted only the teachers. This intense focus often ignores student actions.

Classroom Visit Types

There are various ways to collect data from classrooms during instruction. Recognize, however, that classroom visit data are informal, summative data. Results are provided to teachers after the fact. Several factors highlight classroom visits for monitoring implementation. These factors include the following:

- Collected data are focused on the Practices.
- Collected data are "snapshots" of multiple classrooms over time.
- Collected data do not record individual teacher names.
- Classroom visits are roughly three to five minutes per classroom.
- Tally sheets are not marked within the classroom.

Abiding by these factors is important for building relationships and trust. From these factors, several types of classroom visits may be used.

Classroom visits are used on a regular basis to monitor, in this circumstance, student actions. Single individuals may conduct visits. The visitor enters a classroom, watches the students, moves around the room if possible, and leaves within three to five minutes. Once outside the classroom, the visitor records the time and what was seen on a Classroom Visit Tally—as shown in Figure 9.1. The visitor quickly moves to another classroom and repeats the process. The goal is to visit as many classrooms as possible within a set time. The same classroom should, if possible, be visited more than once.

Teams may also conduct classroom visits. Teams follow the same procedures listed in the previous paragraph, only a classroom is visited by different people at different times during the lesson. If desired, when working with teams, two visitors may work as partners when visiting.

Schools may exchange teams to conduct classroom visits. This opportunity may provide a different perspective for the people engaged in the classroom visits as well as the ones receiving the data.

Schools may employ outside consultants to conduct classroom visits. Consultants familiar with the Common Core Practices may either collect data independently or work with leadership teams to collect data. Again, this opportunity can provide a different perspective of classroom instruction.

Classroom Visit Tally—Students

Collecting data by conducting classroom visits (walk-throughs) is only useful if the collected data are used as feedback to inform teachers about their instructional practices. In this case, teachers are receiving data concerning the actions of their students. Once the data are collected, leadership teams compile the data gathered from the Classroom Visit Tally—Students (Figure 9.1).

The Classroom Visit Tally identifies actions students are taking if they are engaged in the CCSS Practices and the proficiency indicators from the Matrix. Visitors are directed to enter the classroom and, as unobtrusively as possible, carefully watch what students are doing. If not disruptive, visitors may circulate around the room observing and listening to students. After three to five minutes, the visitors exit the classroom, record the time, and then record what they saw on the Classroom Visit Tally. Visitors must only record specific actions in which they personally observed the students engaged. If no student actions were witnessed, none are recorded. Visitors, in these circumstances, make a note that classrooms were visited. Finally, visitors record no information that might disclose the name of the teacher. In large schools, the grade level or subject may be noted for more specific analysis.

Collected data are recorded on the Classroom Visit Tally. At this point, only student data are recorded on the sheet.

Teacher Self-Assessment of Student Actions

In conjunction with classroom visits, teachers should independently reflect on the student actions occurring within their classrooms. These are the same actions used in the classroom visit and are directly related to the Standards for Mathematical Practice. Teachers are reflecting on the actions students are taking within the classrooms and the levels of proficiency (Initial, Moderate, or Successful) students are demonstrating. Teachers use a form (Figure 9.2) to monitor their progress in moving students toward the Advanced level on the Matrix. The only identifier on the form is the date.

Teachers' names are not wanted or needed on the form. These self-assessment forms may be included in the discussions concerning

Figure 9.1 Classroom Visit Tally

School: _____ Grade or course: _____

Date: _____

Students are	Classroom				
	CR1	CR2	CR3	CR4	CR5
Working in pairs					
Engaging in appropriate discussions					
Using models or pictures					
Explaining their thinking					
Organizing their thinking					
Working in small groups					
Working enthusiastically					
Using mathematical tools					
Discussing others' thinking					
Seeking alternate solutions in groups					
Using correct vocabulary					
Demonstrating persistence					
Discussing others' solutions					
Finding and using patterns					

Comments:

Classroom 1: _____

_____

Classroom 2: _____

_____

Classroom 3: _____

_____

Classroom 4: _____

_____

Classroom 5: _____

_____

Source: © 2008 Hull, Balka, and Harbin Miles. Adapted from Hull, Balka, and Harbin Miles (2012).

Figure 9.2 Teacher Self-Assessment of Student Actions

Action: Students are working in pairs.			
Degree	**Initial**	**Moderate**	**Successful**
Several times daily			
Several times weekly			
Seldom			
Not used			
Action: Students are engaging in appropriate discussions.			
Degree	**Initial**	**Moderate**	**Successful**
Several times daily			
Several times weekly			
Seldom			
Not used			
Action: Students are using models or pictures.			
Degree	**Initial**	**Moderate**	**Successful**
Several times daily			
Several times weekly			
Seldom			
Not used			
Action: Students are explaining their thinking.			
Degree	**Initial**	**Moderate**	**Successful**
Several times daily			
Several times weekly			
Seldom			
Not used			
Action: Students are organizing their thinking.			
Degree	**Initial**	**Moderate**	**Successful**
Several times daily			
Several times weekly			
Seldom			
Not used			

(Continued)

Figure 9.2 (Continued)

Action: Students are working in small groups.			
Degree	**Initial**	**Moderate**	**Successful**
Several times daily			
Several times weekly			
Seldom			
Not used			

Action: Students are working enthusiastically.			
Degree	**Initial**	**Moderate**	**Successful**
Several times daily			
Several times weekly			
Seldom			
Not used			

Action: Students are using mathematical tools.			
Degree	**Initial**	**Moderate**	**Successful**
Several times daily			
Several times weekly			
Seldom			
Not used			

Action: Students are discussing others' thinking.			
Degree	**Initial**	**Moderate**	**Successful**
Several times daily			
Several times weekly			
Seldom			
Not used			

Action: Students are seeking alternate solutions in groups.			
Degree	**Initial**	**Moderate**	**Successful**
Several times daily			
Several times weekly			
Seldom			
Not used			

Action: Students are using correct vocabulary.

Degree	Initial	Moderate	Successful
Several times daily			
Several times weekly			
Seldom			
Not used			

Action: Students are demonstrating persistence.

Degree	Initial	Moderate	Successful
Several times daily			
Several times weekly			
Seldom			
Not used			

Action: Students are discussing others' solutions.

Degree	Initial	Moderate	Successful
Several times daily			
Several times weekly			
Seldom			
Not used			

Action: Students are finding and using patterns.

Degree	Initial	Moderate	Successful
Several times daily			
Several times weekly			
Seldom			
Not used			

Note: The data on this form are never used by leaders to confront teachers. If there is a discrepancy between the Classroom Visit Tally form and the Self-Assessment form, then conversations take place to clarify the actions.

classroom visit data. Copies of the forms are brought to the meeting, combined, and shuffled. Another possibility is that a master form is created prior to the meeting. Any willing teacher, a coach, or specialist may do this, provided copies of the form are collected and given to the individual prior to the meeting.

The data from the forms are never used by any leader to confront the teachers. If there is a discrepancy between the visits and the Self-Assessment, then conversations take place to clarify the actions. What are teachers using to identify particular actions, and what are visitors using to identify the same actions?

Action: Students are explaining their thinking.			
Degree	Initial	Moderate	Successful
Several times daily			
Several times weekly			
Seldom			
Not used			

Math Coach Scenario

Mary S. is a middle school mathematics coach. Before being assigned to the position three years ago, Mary taught for seven years. She is assigned to several schools and feels she has good rapport with the teachers.

During her three years as a specialist, Mary has been able to develop a very positive working relationship with the teachers. She is often in classrooms and frequently helps with planning and presenting lessons. She feels the teachers at the school are all very good. She certainly knows they work hard. Mary also strongly believes everyone should strive to improve, including herself.

As the Common Core State Standards began reaching her school, she carefully read through the documents. She knew she would be part of a team asked to align the new content. When she first read over the Standards for Mathematical Practice, she felt reassured. The wording was perhaps different, but these ideas were the same ones she and the teachers aspired to accomplish every day.

Mary was given the opportunity to attend a regional mathematics meeting being held near her city. She was glad to attend, and she sought out sessions concerning the Common Core. In one session, she was given a Practices Proficiency Matrix. During the session, she worked various problem types and compared the thinking required by the tasks to the practice indicators. She felt her first shadow of doubt.

Mary realized that she believed teachers were implementing new strategies that increased student engagement, but she really had no evidence. Furthermore, she was not sure what individual teachers were actually doing on a daily basis. Were the students discussing mathematics? Were they collaborating? Were most lessons still procedural in nature? She needed to find out.

Mary decided to institute classroom visits. She knew she needed permission from her principal, and she needed to clear this with the teachers. The principal turned out to be easy. She listened to Mary, asked what Mary hoped to accomplish, cautioned her to avoid anything that appeared to be evaluative, and asked her to get the teachers' okay.

Visiting with the teachers was far more difficult than Mary envisioned. When Mary explained what she wanted to do, the teachers began asking lots of questions:

> Why do you need to do this?
> What do you think we are doing wrong?
> What's wrong with how we have been working?
> What happens to the data and reports? Who sees them?
> What can you see in just a few minutes?

Mary found herself becoming very defensive, before she realized that the questions were legitimate concerns, and mostly ones she had not thought about. She told the teachers that their questions raised some very important issues and that the teachers needed to be part of the process. Mary asked that the teachers please continue to work with her, and she asked if one of the teachers would also collect data if Mary could get the principal's approval.

In subsequent meetings, the teachers carefully analyzed the form. They also named a teacher to help Mary conduct the visits, since Mary had received the principal's permission. The teachers agreed to give the classroom visits a try. (Scenario continued in Chapter 10.)

Having Productive Conversations

1. Envision a mathematics classroom engaged in the Practices. What might a visitor reasonably see in three to five minutes?
2. Why is the time for classroom visits limited to three to five minutes? What issues does this avoid?
3. How can classroom visits support change initiatives?
4. Who is available to conduct classroom visits?
5. How can teachers be used to conduct classroom visits?
6. What training or support can be offered between visits?
7. What additional ways can classrooms be visited that would support change, yet remain nonevaluative?

10

Solution Step 2

*Using Classroom Visit
Data—Assessment of Student Actions*

To change classroom behaviors, leaders and leadership teams must monitor and support teachers as they acquire new skills and transform their instruction. This requires classroom visits that are nonevaluative and include professional collaboration, professional learning, coaching, and consistency in actions.

After the data on student actions from classroom visits are collected, feedback to teachers is vital. First, teachers need to see the collected data so they are assured the data are not evaluative, and they can better understand and participate in the process. Second, teachers need to engage in conversations about what the data reveal in relation to students' proficiency on the CCS Standards for Mathematical Practices. These conversations must be positive and lead to shifts in thinking about student learning, the Practices, and instructional strategies that increase student actions.

Although schools usually collect data from multiple sources, such as grades, state assessments, or benchmark assessments, these data sources fail to reveal what students were actually doing in classrooms during instruction. For this reason, we stress that data are collected specifically about students and the classroom actions they were engaged in during instruction.

Conducting Productive Conversations

The term "productive conversations" has two key components. First, conversations mean an exchange of information. These conversations relate to an important topic. The "productive" part means there are specific, positive outcomes from the exchange of ideas. Ironically, productive conversations, while positive, may best be described by what they are not. Productive conversations are not

- Presentations,
- Fault-finding opportunities,
- Directive,
- Defensive, or
- Confrontational.

Productive conversations are based on an analysis of the data and then a reflection on the data analysis. Analysis means looking at the data to determine what was noted and what was not. Analysis avoids opinions as much as possible. Teachers want to clearly identify what the data indicate. For instance, questions such as the following may be discussed.

- How many classrooms were visited?
- What span of time did the visits cover?
- Which indicators were noted?
- How frequently were they noted?
- Which indicators were not noted?

If there is previous data, then,

- How does this data compare to previous data?
- What changes occurred?

Analysis sets the positive, nonblaming tone of the conversations. Everyone should participate and offer their thoughts. Collaboration is very important.

Once a careful analysis has been conducted, productive conversations turn to reflection. Reflection includes the following:

- How often should the indicator be seen?
- What is an acceptable balance of indicators?
- Which student actions are most directly related to the indicators?
- What does a classroom usually look like?
- What are students doing regularly?
- What are students apparently not doing?
- What is the next step to more engaging classrooms?

If at all possible, teachers should lead the meeting. Formal leaders may wish to explain the flow of analysis then reflection. Students are the focus, not teachers. Formal leaders may provide support and encouragement and then leave the meeting if they feel they may inhibit conversations. Realistic expectations are very important. Every indicator should not be seen in every lesson. Lesson balance is optimal.

Using the data and conducting productive conversations are intended to result in changing classroom instruction. Change, however, is complicated. As a rule, educational systems have unrealistic expectations concerning change. The current system operations do not support change and actually work to restore the current status. Teachers and leaders need to clarify their understanding of how adults change if the monitoring process and collected data are to achieve the desired results.

Understanding the Change Process

Adults change behaviors for a variety of reasons. They also change at different rates depending on the change being requested. In most cases, teachers begin changing their behaviors when their students demonstrate an immediate, obvious need. This is accomplished through ongoing formative assessment. Summative assessments provide too great a gap between the instruction and the observed need. Teachers and students have moved forward and any reteaching is misplaced.

When teachers are asked to change their instructional strategies, they may or may not see the benefit of such a change. Moreover, they are not yet proficient in the new strategy, nor are the students accustomed to the change. Teachers are asked to take a great leap of faith when told to use an instructional strategy they have not practiced, are unsure what the strategy looks like in action, and are unsure of the outcome.

This skepticism or reluctance to participate in a new initiative is a product of current system thinking and must be changed. Monitoring, without support, will prove frustrating and unproductive. Routinely collecting data around implementation of the Practices works if intervention is taking place in the interim. According to Hall and Hord (2001, p. 82), individuals follow particular levels while learning something new. The levels, from the bottom up, are the following:

- VI Renewal
- V Integration
- IVB Refinement
- IVA Routine
- III Mechanical Use
- II Preparation
- I Orientation
- 0 Nonuse

The levels are rather self-explanatory. In reviewing these levels, it becomes apparent that adults need time to achieve mastery of new strategies. Time alone, however, is insufficient. Time does not assure mastery unless training and support are also included. To reap any benefits from a strategy, Level III is the minimal one a teacher must obtain. In other words, teachers must step forward and attempt to use a strategy after Levels I and II are completed.

This step is difficult for teachers to take, yet a strategy cannot be perfected and "routinized" without this level being undertaken and then surpassed. The current, unsupported, teacher-in-isolation system clearly does not encourage teachers to self-initiate the inclusion of new strategies.

Levels of Adoption

Hall and Hord (2001) identified progressive levels adults pass through as they master a technique. That is, they have reached Level III and are willing to try. In other research, Rogers (1995) helps explain how people decide to take the Level III step. In simplifying Rogers's work, we have identified three groups: initiators, early adopters, and later adopters. These groups form in schools based on the strategy to be adopted, social networks, issues of esteem or respect, beliefs, and outside forces. Each group is moving through the levels of adoption at different times and are, therefore, at different levels with different needs.

The lack of knowledge about the change process is another major system operation barrier. Fundamentally, the system supports everyone getting the same training, in the same way, at the same time, and therefore, everyone is moving to adopt the change in lockstep precision. Nothing could be further from the truth. Even though change initiative after change initiative has failed from this system approach, the belief still persists. Even when the belief may no longer exist in some situations, making necessary changes seems impractical, so the habit, by default, continues. When adopting a new strategy, outside of the classroom professional training only supports Level I: Orientation and Level II: Preparation. Once the strategy is regularly used, outside of the classroom professional learning may influence Level V: Integration and Level VI: Renewal. Nonetheless, the critical Level III: Mechanical use and Level IVA: Routine require hands-on support to be effectively adopted.

Levels of Use	Behaviors Associated With Level of Use
VI Renewal	Explores new and different ways to implement innovation
V Integration	Integrates innovation with other initiatives; does not view it as an add-on; collaborates with others
IVB Refinement	Begins to explore ways for continuous improvement
IVA Routine	Comfortable with innovation and implements it as taught
III Mechanical	Concerned about mechanics of implementation
II Preparation	Begins to plan ways to implement the innovation
I Orientation	Begins to gather information about the innovation
0 Nonuse	No interest shown in the innovation; no action taken

Source: Hall and Hord (2001). www.sde.ct.gov/sde/lib/sde/pdf/curriculum/cali/day2.pdf

Intervention as Support

Earlier, teacher intervention techniques were discussed. Within the discussion, it was noted that teachers must consider intervention by way of ongoing formative assessment as different from intervention through reteaching. Similarly, leaders must consider intervention resulting from classroom visits as different from intervention through corrective measures. When working with teachers to increase student engagement, support and encouragement are needed. What supportive actions are needed to help teachers move from Level II: Preparation to Level III: Mechanical Use? Next, what supportive actions are needed to maintain use of the strategy or technique so teachers increase their proficiency to Level IVA: Routine and higher?

Support comes in a variety of ways. Support may be verbal. Support can come from peers. Support can come from having another person in the classroom to help manage students. Support may come from a person willing to listen. Support may come from someone who has successfully instituted a change. The message is clear. Locate the type of support individual teachers need and provide the support.

Building a Critical Mass

Leaders must think of change as building a critical mass of support, one teacher at a time. Teachers usually adopt change in small groups. Leaders need to know individual teachers in the small group and, if possible, promote collaboration. While first adopting a change and gaining expertise as they move from Mechanical Use to Routine to Refinement and higher (Hall & Hord, 2001), they need the leader's support. As they achieve mastery, they can encourage other teachers to adopt the strategy or change. The leader, through monitoring, is aware of the next individual teacher or small group willing to adopt the change, and the leader is prepared to provide support.

Changing the Culture

As teachers adopt and enact change with leaders' support, both leaders and teachers need to focus on the change and the result of the change. As previously noted, system operations promote stasis. While changes are being made, the forces within the system are working to neutralize the change. These forces are often subtle, but perhaps not. For this reason, leaders and teachers need to guarantee the school culture is shifting with the change.

For example, if teachers are working to adopt the strategy of think, pair-share, then the classroom culture must change also. Students may be unaccustomed to working together. The cultural belief that silence equates with learning must shift. The idea that looking at another student's paper or working together is considered cheating must change. Parental concerns related to fair grading practices may arise. Grading policies may need to be revisited.

Connecting Actions Chart

Students' actions are what drive learning, but teachers' actions drive students' actions. Students are expected to learn mathematics through engaging in the Practices. There certainly is a relationship between students' actions, the Practices, and teachers' actions. This relationship is provided in Table 10.1: Connecting Actions Chart.

The purpose of the chart is to maintain a focus on why students should be engaged in particular actions. In the early stages of implementing a new strategy, such as working in pairs, the focus is on implementing the strategy well. This is normal, but once the strategy is being used effectively, the focus must shift to the outcomes of the strategy. In other words, are students using working in pairs to make sense of problems and the mathematics within the problem?

Table 10.1 Connecting Actions Chart

Students Are	CCSS Math Practices	Teachers Are
Working in Pairs	1A Making Sense of Problems	*Using Think, Pair-Share*
Engaging in Appropriate Discussions	1A Making Sense of Problems 3A Constructing Viable Arguments	*Requiring Thinking*
Using Models or Pictures	2 Reasoning Abstractly 4 Modeling With Mathematics	*Providing Models and Pictures*
Explaining Their Thinking	3A Constructing Viable Arguments	*Asking Thoughtful Questions*
Organizing Their Thinking	4 Modeling Mathematics	*Providing Wait Time*
Working in Small Groups	4 Modeling Mathematics	*Using Grouping*
Working Enthusiastically	1A Making Sense of Problems	*Using Engaging Problems*
Using Mathematical Tools	5 Using Appropriate Tools Strategically	*Providing Mathematical Tools*
Discussing Others' Solutions	3B Critiquing the Reasoning of Others	*Posing Questions to Small Groups*
Seeking Alternative Solutions in Groups	1B Persevering in Solving Problems 3B Critiquing the Reasoning of Others	*Offering Prompts to Small Groups*
Using Correct Vocabulary	6 Attending to Precision	*Supplying Appropriate Vocabulary*
Demonstrating Persistence	1B Persevering in Solving Problems	*Allowing Struggle*
Discussing Others' Thinking	3B Critiquing the Reasoning of Others	*Encouraging Reasoning by Comparing*
Finding and Using Patterns	7 Looking for and Making Use of Structure 8 Looking for and Expressing Regularity in Repeated Reasoning	*Highlighting Patterns*

Teacher actions related to the Practices are discussed in Chapter 11. The Connecting Actions Chart is available for distribution when leadership teams and leaders are ready, but the teachers' actions side should not be completed until after Chapter 11 is studied.

Math Coach Scenario (Continued)

After working out the details with the teachers, a decision was made that classroom visits should take place for shorter segments of the day for several days rather than multiple times during a single day. The principal liked this plan because she could arrange to cover the participating teacher's, Sara's class, rather than hire an all-day substitute.

On day 1, Mary and Sara had an hour scheduled to visit classrooms. There are 12 mathematics teachers in the school. Mary and Sara decided to work together as a way to check for agreement in what was happening in the classrooms. They also agreed that only one form would be submitted.

Mary and Sara entered a classroom. Mary was named as timekeeper. They watched two students for five minutes. Mary and Sara stepped into the hall and independently tallied their form. They repeated this process five times. Because this process was new and classes were not always in the next classroom, Mary and Sara only had 15 minutes before the hour was up. They agreed to meet after school to compare sheets.

In the follow-up meeting, Mary and Sara worked out their differences and reached accord on what various indicators meant.

On Day 2, Mary and Sara again had an hour, but it was a different hour of the day than the first day. Mary and Sara independently visited rooms. They organized their room list so no one teacher was visited more than twice. This process was repeated for Day 3, and again, a different hour was selected. At the end of Day 3, Mary and Sara visited 39 classrooms. In their final tally, several teachers were visited twice.

Mary and Sara met after school to record data on the Classroom Visit Tally—Students form (Figure 9.1). After completing the tally, copies were made of the sheet, and then all forms were sealed in a large envelope. The teacher meeting was scheduled for the following afternoon.

Mary and Sara stared at the totals. They were stunned, then concerned, and then very nervous. The upcoming meeting was not going to be pleasant. They discussed canceling the meeting and abandoning the entire process of classroom visits. Yet Sara wished to press forward. Although truly nervous, Sara asked Mary to let her conduct the meeting. After further conversation, Mary reluctantly agreed.

At the start of the meeting, Mary explained the process used to collect data. She then told the teachers that she and Sara had recorded the totals. Mary opened the sealed envelope in front of the teachers and then shredded the individual forms in front of the teachers.

Sara took a deep breath, and said, "To be honest, I am really nervous. Mary and I discussed giving up on the process, but I think you deserve the information. Just as a reminder, Mary and I visited all of you more than once over several days, and we were able to visit 39 classrooms. We never stayed longer than five minutes, and we only watched for student actions on the list. I have a copy of the totals for each of you. When I hand you a copy, please look it over by yourself for a few minutes. I would like you to think about your reaction, and then formulate your questions."

Sara handed out the totals from the Classroom Visit Tally—Students form, and the room was very quiet for several minutes. The totals for observed actions were the following:

Classroom Visit Tally—Students	
Working in pairs	4
Engaging in appropriate discussions	1
Using models or pictures	8
Using mathematical tools	1

Finally, one teacher said, "I just don't believe this. Had you stayed in my room, you would have seen almost every one of these actions." There were general murmurs of agreement, and then the room erupted into multiple conversations with everyone talking.

When there was a lull in the talking, Sara called the teachers back together and said, "You are probably correct. Remember, this data is not about any one person at any one time. Mary and I visited 39 classrooms over several days. We were present for the beginning, middle, and end of various lessons. This is what we saw."

In the silence, Mary asked, "How many times do you think we should have seen students working in pairs? How often do you use the strategy?"

After some discussion, the agreement was that teachers used the strategy three or four times per period. Mary said, "That number sounds reasonable. So in 45 minutes we would see the strategy about three times, or once every 15 minutes. Since we were in classrooms only five minutes, we would have seen the strategy about a third of the time it was used. Think about this, and see if you agree."

The teachers thought for a few minutes and talked quietly among themselves. Finally, one teacher said, "This sounds about right, but maybe fewer times." Mary said, "If we stay with the three, then one-third of 39 is 13. On average, if the action is used three times per class, then we should have recorded this action at least 10 times. We recorded four. What do you think is happening?"

Mary told the teachers that perhaps the meeting should end today so everyone would have time to think things over and talk to one another. She assured the teachers that no names were attached to any of the data and that both she and Sara would be available. The next meeting was scheduled for two days later.

Meeting 2

There was certainly a buzz within the building for the next two days. Several teachers privately talked to Sara and Mary. At the start of the meeting, Mary explained again why she felt classroom visits were needed. She reminded the teachers that everyone in the room cared deeply for their students, and the entire process was designed to help students learn mathematics. She handed out copies of the Practice Proficiency Matrix again and said, "This is what we all want for our students. I need to hear and understand what you are thinking."

There was an uncomfortable silence, but neither Sara nor Mary spoke. Finally, a teacher, Elizabeth, spoke up. Elizabeth said, "You both visited my classroom several times. I really don't remember if I ever had the students working in pairs when you were there. I do know that I don't use this strategy as often as I should. No, that's not right either. I haven't been letting students work in pairs except maybe when they are doing homework at the end of the period. For me, these numbers on the form are correct. I know this sounds silly, but I just get so busy teaching that I forget to stop, and then the class is over. There is always so much to do. I need some suggestions."

Sara said, "Elizabeth, thank you. Now, I have a confession, too. If my classroom had been visited, the numbers for having students work in pairs would have been even lower." Everyone in the room laughed.

Sara continued, "I know we can do this, and I believe we should do it. Our students need us to lead the way. So what do you say we figure out how to get things moving?" (End of scenario.)

Having Productive Conversations

1. Data are neither positive nor negative. What does this mean?
2. How can data be used to inform instruction?
3. How does classroom visit data relate to the Matrix?
4. How can data indicate additional support is needed?
5. How can data indicate additional strategies can be incorporated?

11

Solution Step 3

*Monitoring Teacher
Actions Related to the Practices*

Leaders must garner support from individuals responsible for implementing the change—teachers. Leaders have the ability to make things happen. They work to arrange meetings and focus conversations. Leaders are major players in establishing a school vision and mission. They have access to various types of student performance data, as well as information from their state concerning assessment.

While leaders have access to classrooms, and can open doors, they must rely on teachers to actually make a difference for students. Further, teachers need support from both leaders and other teachers to make significant change. As a result, leaders must empower teachers. Collaboration across all levels of the school system is critical.

If leaders are going to support teachers as they work to change their instructional practices, then leaders must know what actions teachers are taking. Leaders, including teachers, have entered classrooms to collect data concerning student actions around the Practices. Now, leaders, again including teachers, must collect information related to the actions teachers are taking. Failure to monitor the implementation of the Practices remains a major obstacle to success. Figure 11.1 lists specific teacher actions that directly relate to the student actions and Practices.

In preparing to add another dimension to monitoring, serious conversations need to occur between leaders and teachers. While the student actions, teacher actions, and Practices are displayed on the previously offered Table 10.1, teacher actions have not been documented, only presented. Now, there are conversations aimed at discussing the types of actions students are taking and the actions teachers are taking. The teacher actions, as with student actions, are based on a combination of the Standards for Mathematical Practice and the Strategy Sequence.

Figure 11.1 Classroom Visit Tally—Teacher

School: _____ Grade or course: _____

Date: _____

Teachers are	Classroom				
	CR1	CR2	CR3	CR4	CR5
Using think, pair-share					
Requiring thinking					
Providing models and pictures					
Asking questions					
Providing wait time					
Using grouping					
Using engaging problems					
Providing mathematical tools					
Posing questions to groups					
Offering prompts to small groups					
Supplying appropriate vocabulary					
Allowing struggle					
Encouraging reasoning by comparing					
Highlighting patterns					

Comments:

Classroom 1: _____

_____

Classroom 2: _____

_____

Classroom 3: _____

_____

Classroom 4: _____

_____

Classroom 5: _____

_____

Source: © 2008 Hull, Balka, and Harbin Miles. Adapted from Hull, Balka, and Harbin Miles (2012).

Visitors to classrooms who are collecting data on the Classroom Visit Tally sheets about teachers' actions quickly realize that some teacher actions do not appear on the form. This absence of certain teacher actions might raise concerns from the visitor. Not all actions by teachers are directly related to the Standards for Mathematical Practice. Actions such as teacher-directed discussion, teacher lectures, teacher modeling, student independent practice, and individual student recall are not present on the form.

This absence does not mean that these actions are ineffective or should not ever be used. The point, however, is that these actions do not generally reflect the Practices. Moreover, if these are the only instructional strategies used, then the Practices are not being implemented.

As a final note, when change is first initiated and classroom visits are under way, it is very likely that visitors will not see many teacher actions that support the Practices. Logically, this makes sense. Difficulties arise, however, when visitors are painfully aware of the absence of any indicators being recorded on the form and feel obligated to record something. This is a serious mistake, and it undermines the entire monitoring process. Visitors must abide by one inviolate rule—if the visitor does not witness the action, the action is not recorded.

The form is used as a point to initiate conversations. Leaders and teachers must realize that every action will not be seen during a classroom visit, nor are all the actions adopted at early stages of change. Teachers and leaders should select several of the actions they feel are in use and are reasonable. These are the actions that visitors should be attuned to seeing.

Using the Classroom Visit Tally—Teachers Form

In the same way the Classroom Visit Tally—Students form was used, classroom visitors want to enter classrooms for a brief time. After three to five minutes, visitors leave the room, record the time, and record what they saw. Again, the form does not make note of the teacher's name. Information on a single visit is not valuable. Only when sufficient classrooms have been visited to identify patterns or trends is the data worthwhile.

The data are compiled and may be used in conjunction with obtained student visit data or just the teacher data. The student visit actions and teacher visit actions are in one-to-one correspondence. This correspondence is easily recognized when the forms are compared. The focus of the conversations is on degree of implementation and usage of the actions within the mathematics department, grade level, or subject area.

Conversations About the Data

Leaders and teachers must have open, honest dialogue concerning the data generated. The importance of the information is to capture what is happening and what needs to happen less frequently or more frequently. If student data are also used, then conversations are directed at the relationship between the two sources.

Discussion questions may include the following:

- What is the most frequently noted teacher action?
- What teacher actions are less frequently noted?
- Are some actions not recorded?
- Are some of the not-witnessed actions being used?
- Is there compatibility between student actions and teacher actions?
- What teacher actions should be increased?
- If certain teacher actions are increased, what student actions should also increase?

These questions are nonjudgmental and are designed to promote effective instructional decisions that support the Practices.

Working on Individual Needs

Classroom Visit Tallies are important tools that support effective change. The focus for this data is on large numbers of visits, and not associated with individual teachers. However, at some point, individual teachers must be the focus. Doing so is a critical shift and must be handled wisely. Even though this is an important piece to change, we strongly **discourage** the use of the Classroom Visit Tally forms for recording individual observation data.

Mathematics leaders are working with teachers on a regular basis. After several planning and coteaching opportunities and numerous conversations concerning student and teacher actions, both the mathematics leader and teacher have a very good idea of what actions are being used in individual teachers' classrooms.

Further information can be gathered. Teachers can complete the Teacher Self-Assessment for Personal Actions (Figure 11.2) as a way to decide what actions should be taken next. This form can be completed by the teacher alone or in a conversation with the mathematics leader.

Figure 11.2 Teacher Self-Assessment for Personal Actions

Action: Using Think, Pair Share			
Degree	**Initial**	**Moderate**	**Successful**
Several times daily			
Several times weekly			
Seldom			
Not used			
Action: Requiring Thinking			
Degree	**Initial**	**Moderate**	**Successful**
Several times daily			
Several times weekly			
Seldom			
Not used			
Action: Providing Models and Pictures			
Degree	**Initial**	**Moderate**	**Successful**
Several times daily			
Several times weekly			
Seldom			
Not used			
Action: Asking Questions			
Degree	**Initial**	**Moderate**	**Successful**
Several times daily			
Several times weekly			
Seldom			
Not used			
Action: Providing Wait Time			
Degree	**Initial**	**Moderate**	**Successful**
Several times daily			
Several times weekly			
Seldom			
Not used			

Action: Using Grouping

Degree	Initial	Moderate	Successful
Several times daily			
Several times weekly			
Seldom			
Not used			

Action: Using Engaging Problems

Degree	Initial	Moderate	Successful
Several times daily			
Several times weekly			
Seldom			
Not used			

Action: Providing Mathematical Tools

Degree	Initial	Moderate	Successful
Several times daily			
Several times weekly			
Seldom			
Not used			

Action: Posing Questions to Groups

Degree	Initial	Moderate	Successful
Several times daily			
Several times weekly			
Seldom			
Not used			

Action: Offering Prompts to Small Groups

Degree	Initial	Moderate	Successful
Several times daily			
Several times weekly			
Seldom			
Not used			

(Continued)

Figure 11.2 (Continued)

Action: Supplying Appropriate Vocabulary			
Degree	**Initial**	**Moderate**	**Successful**
Several times daily			
Several times weekly			
Seldom			
Not used			

Action: Allowing Struggle			
Degree	**Initial**	**Moderate**	**Successful**
Several times daily			
Several times weekly			
Seldom			
Not used			

Action: Encouraging Reasoning by Comparing			
Degree	**Initial**	**Moderate**	**Successful**
Several times daily			
Several times weekly			
Seldom			
Not used			

Action: Highlighting Patterns			
Degree	**Initial**	**Moderate**	**Successful**
Several times daily			
Several times weekly			
Seldom			
Not used			

Note: The data on this form are never used by leaders to confront teachers. If there is a discrepancy between the Classroom Visit Tally form and the Self-Assessment form, then conversations take place to clarify the actions.

Experimenting, Using, Integrating

Teachers are familiar with the form since they have been completing a similar form without their name as data to compare to the Classroom Visit Tally. The Self-Assessment asks for more information to help teachers and leaders select the next steps in change.

Mathematics Collaborative Log

With the data collected, mathematics leaders need to support individual teachers as they continue to make progress. In private conversations, mathematics leaders may find the Mathematics Collaborative Log (Figure 11.3) form to be useful. This form focuses on specific actions for a teacher to take and leaders to take.

The form provides a framework for leading conversations concerning continual progress and the shared work between teachers and leaders. Conversations include areas that are working as well as challenges. Leaders and teachers decide on the next steps needed and a date for taking the next steps.

Teacher Planning Guide

One additional form is available for use by teachers and, perhaps, by leaders. The Teacher Planning Guide (Figure 11.4) is used to help teachers plan lessons related to the Matrix. The purpose of the form is not to "fill in" the form, but rather think about upcoming lessons and significant issues that must be addressed. The questions help to guide thinking regarding the students.

Having Productive Conversations

1. What is the purpose of classroom visits?
2. How can classroom visit data inform instruction?
3. How can teachers and leaders collaborate effectively with visit data?
4. What can leaders do to ensure confidentiality?
5. How can classroom visits promote the Practices?

Figure 11.3 Mathematics Collaborative Log

Teacher _____ Grade _____ Coach _____ RHM _____ Today's Meeting Date _____

What Is Working	Challenges/Concerns

Teacher's Next Steps	Coach's Next Steps

_____ Reflecting on smart goals	_____ Observing student learning	_____ Coplanning lessons	
_____ Analyzing student data or student work	_____ Sharing effective instructional strategies	_____ Providing resources	
_____ Teaching the curriculum/discussing SOLs	_____ Sharing models/manipulatives	_____ Reflecting — post conference	
_____ Observing teacher instruction	_____ Using technology to *teach* math	_____ Other	

Next Meeting Date _____

Figure 11.4 Teacher Planning Guide

What do students need to know? (CCSS)	**What materials and resources do I have?**
What do students need to do? (Matrix)	**What do I need to do?**
What are common misconceptions or errors?	**How will I assess their learning?** Formative Summative

Lesson Flow

Setting the Stage
Exploring
Summarizing

12

Solution Step 4

Gathering and Using Additional Data

There are multiple sources of data for school personnel to use. Classroom visit data are essential and may be gathered in several ways other than the Classroom Visit Tally form. Still, other data are available. Leaders and leadership teams need to routinely analyze data to ensure the students are making progress. Data, for purposes of this book, are used to inform instructional decisions.

Assessments Collectively

All forms of assessment can have a useful purpose and provide worthwhile information if the data collected are used. In fact, once classroom ongoing formative assessments are enacted, formal and informal assessments gain more usefulness. With the daily use of formative assessments, formal and informal assessments serve to affirm that instructional changes are working as intended and that overall progress is being made. In addition, formal and informal assessments provide clues to teachers concerning students' misconceptions that teachers need to carefully monitor. Rigor cannot be attained without effective use of assessments.

Achievement Data

Students are routinely tested in mathematics classrooms. Leaders and leadership teams need to selectively use particular data to validate that changes in instruction also result in positive changes in student learning for all students. With classroom visits, leadership teams recognize whether instruction is shifting. As evidence mounts that strategies are used more frequently and more effectively, leadership teams need to seek validation of teachers' efforts by locating data related to student learning.

Teachers may give common assessments to students and then compare student work with an emphasis on student understanding. Teachers may give common six- or nine-week exams that can be analyzed. Teachers may teach common instructional units that include challenging problems. Student work on these problems may be compared.

Leaders, and leadership teams, may wish to use some achievement data to regularly highlight gains. For instance, if common six-week tests are given, then a ratio of students passing to students failing can be collected. This ratio is charted and compared every six weeks. Many schools require regularly scheduled benchmark assessments. These assessments can also be recorded as a pass/fail ratio.

Regardless of the data source and method of recording, student achievement data must be compared to data being collected during classroom visits. For instance, leaders notice a definite increase in the strategy of think, pair-share. What data are available to indicate that with the increase in the use of the strategy there is an increase in student learning? As a note, at this level of analysis, the relationship is correlational, not causal.

Data From Classrooms

There are many reasons for classroom doors to be opened wide. Mathematics instruction has little chance of rising to the challenge of the Practices if teachers quietly work in isolated conditions. As teachers and leaders grow more comfortable with data gathering, monitoring, and feedback, additional methods for collecting data concerning both student actions and teacher actions may be incorporated. As always, leaders responsible for evaluations need to be cautious about the types of visits they use and the data they gather. We offer multiple ways for data to be collected.

Specified Classroom Visits

Frequently, leaders only have a short time available. They can use this time effectively with only a piece of scratch paper. In this technique, leaders are looking for one or two student actions. As an example, leaders may elect to look for one action—working in pairs. Leaders walk around the classrooms very quickly and record "yes" or "no." They can easily establish a ratio of the presence of the observed strategy to the total number of classroom visits.

The important point is that monitoring and feedback must be routine actions of leaders and leadership team members. Improvement efforts reside in monitoring progress.

Additional specified classroom visits can be used, but more time is needed. In the next two techniques, 8 to 10 minutes are recommended. In Technique 1, leaders are again looking for one or two student actions, but they are also looking for the related teacher actions. For example, leaders may look for students sharing their thinking. Again, the results are yes or no, but if a yes is noted, then what are the strategies teachers are using to promote students sharing their thinking?

In the second specific visit technique, leaders look for quality. Again, the leader is noting either yes or no, but if yes, the leader remains longer in the classroom to assess the quality of the strategy. For example, leaders note that students are working in pairs, but they are doing their homework and only sharing answers as compared to seeking other solutions to a challenging problem.

Validity Visits

As teachers and leaders become more practiced with collecting data, they can become lulled into "seeing" actions that are not present. In their experience, they know the teacher uses the strategy, so they document reference to the strategy as if they actually observed use of the strategy. There is also the possibility that the strategy has been gradually diluted. Therefore, validity visits are very helpful.

In validity visits, a trusted colleague who is familiar with the processes and Standards goes with the leader. Each person independently records what he or she saw. After several visits, the two people have a private meeting to compare and contrast the tallies.

Another validity visit option for leaders is to conduct the visits with two teachers. The same procedures are followed with individuals independently recording their tallies, and then comparing the findings. This method provides an excellent way to affirm expectations and clarify the message. The third option is for a consultant or outside person to check for validity.

Reverse Visits

In a reverse visit, visitors are not using a form during their visits. They work in pairs and stay in a classroom for about 15 minutes. One of the visitors writes down as much as possible about the students' statements and actions. The other visitor writes down as much as possible of the teacher's statements and actions. After completing their visits of the designated number of classrooms, the visitors compare their scripts to either the Classroom Visit Tally form or the Practice Matrix form.

Reverse visits provide an opportunity to look at classroom instruction independent of the developed forms. The information collected may help direct future professional development plans or identify instructional areas that need to be encouraged or discouraged. Reverse visits also assist in reviewing the Classroom Visit Tally forms. Are the actions clearly indicating what is desired in the classroom?

Teacher-Requested Visits

Teachers should always feel they may request visits to support their Practice. In teacher-requested visits, the teachers determine the format and time for the visit. Teachers may want a general overview of lesson effectiveness or a respected opinion concerning the use of selected strategies. Visitors conducting teacher-requested visits should adhere to the bounds identified by the teachers. Teacher-requested visits are nonevaluative and no documentation should leave the teachers' hands.

Supporting Teachers' Change Efforts

To support teachers as they work to shift their classroom strategies toward those identified by the Practices, leaders must actually be in classrooms on a regular basis for the express purpose of supporting teachers. In our book, we have identified multiple ways for leaders to accomplish this. The ways recommended still work, but the focus for the leader shifts.

In understanding change, leaders, as well as teachers, are aware that perfecting a strategy is not an overnight phenomenon. Learning to use a strategy effectively requires time, practice, feedback, and training. In a way to simplify the levels of change identified by Hall and Hord (2001), three terms are selected for the degree of use: experimenting, using, and integrating.

Experimenting

When first attempting to incorporate a new strategy, teachers tend to overregulate the conditions for using the strategy. The keywords are *awkward* and *controlling*. This is normal. Experimenting means the strategy is being tested within classroom routines and introduced to the students. A strategy, such as think, pair-share, may be reasonably successful when first used, but again, it may not be as successful as desired. Again, this is normal. Teachers and their students must work through a degree of experimenting. Each use of the strategy increases the benefits and the ease of application.

Using

There is no defining line between experimenting and using a new strategy. The keywords are *confident* and *flowing*. Using the strategy is achieved when there is far more confidence and comfort obtained. In other words, teachers no longer need to control every aspect of the strategy but find they employ the strategy when they observe the need.

Integrating

A strategy that is at the integrating degree means it is now a permanent fixture in the classroom instructional routine. The keywords are *intentional* and *transitioning*. Teachers consciously decide to apply the strategy for specific, anticipated outcomes from students. When thinking about their upcoming lesson, teachers know when and how the strategy should be used. Also, the strategy usage is flexible. Teachers use the strategy during formative assessment to increase learning and decrease misunderstandings.

Adoption Stages

Because teachers adopt change in small groups based on individual motivations and interest, then at any point in time, different teachers are at different points in the change process. Generally, this means some teachers are still nonusers, some are experimenting with a strategy, some are using the strategy, and others are integrating the strategy. As a result, if leaders wish to successfully implement a strategy, they must know which teachers are at what point, and what training support they need. Teachers must be aware of this same information.

One other critical observation needs to be considered. Leaders follow the same change process as teachers. When leaders first enter classrooms for visit data, they are at an experimenting degree. Leaders must persist in their efforts to improve their skills. This learning curve extends to professional conversations based on the visits. Often, leaders and teachers are both at an experimenting degree for these early dialogues. Teachers and leaders must be patient, forgiving, and open. The focus is always on what is best for the students.

Documenting Progress

Leaders are obtaining much data concerning classroom actions by students and by teachers. Some of the data is teacher specific, and some is not. The reason trust must be established is because leaders now start identifying teachers by name. The purpose of classroom visits adds another dimension. (As a note, this step does not replace classroom visits for implementation.) This dimension assists leaders, as well as teachers, in recognizing where they are in the entire change process. Figure 12.1 is used to record the data.

Leaders working directly with teachers to support implementation of the Practices need to organize their thoughts, formulate some plan of action, and rely on some evidence. The Teacher Implementation Progression offers a way to get better organized and, thus, more productive. The form also provides opportunities for more depth in individual professional conversations with teachers.

The form is for leaders' use only. We do not recommend the completed form be distributed. While it is nonevaluative since no performance criteria are identified, the names make the list confidential. There is a careful balance required between teachers' need to know about the form and how the information is used. Teachers must know about their efforts to change and have a realistic understanding of the progress they have made. However, teachers must also acknowledge that collected information regarding other teachers is confidential unless the teacher decides to share the information with a group.

Completing the Form

Data for the form may be collected in several ways. Since leaders are routinely in classrooms, the form may be completed by professional opinion. In other cases, classroom visit data can be used to directly look for information contained in the form. Leaders should also freely gather teachers' opinions about their thoughts concerning strategy inclusion. While the form with teachers' names is confidential, the categories on the form should be very public. Teachers are already discussing their

Figure 12.1 Teacher Implementation Progression

Teacher		Grade/Course		
Strategy	Description	Experimenting (controlled conditions, awkward)	Using (confident, flow)	Integrating (intentional use, transition)
Initiating think, pair-share	• In pair-share, teachers ask a question or assign a problem and allow students to think and work with a partner for one to three minutes before requesting an answer to the question or problem. • In think, pair-share students are given a brief period to think independently before working with a partner.			
Showing thinking in classrooms	• Every pupil response (EPR) strategies include such responses as "thumbs up/thumbs down," or use of individual whiteboards for noting answers. • Students are pressed to be more aware of their thinking and express their thinking in more detail. • To reveal student thinking, more challenging, open-ended problems are needed.			
Questioning and wait time	• As thinking is increased in the mathematics classroom, better questioning and wait time are required. • Teachers provide thought-provoking questions to students, and then allow the students time to think and work toward an answer.			

	Empowerment Strategies			
Grouping and engaging problems	• Students are given challenging problems to work, and allowed to work on the problem in groups of two, three, or four. • Challenging mathematics problems take time, effort, reasoning, and thinking to solve.			
Using questions and prompts with groups	• As students are provided with opportunities to solve challenging problems in groups, teachers need to increase their ability to ask supporting questions that encourage students to continue working, provide hints or cues without giving students the answers, and ask probing questions to better assess student thinking and current understanding.			
Allowing students to struggle	• Appropriate degree of difficulty is foremost on teachers' minds. If the problem is too easy, students do not need to struggle. If the problem is far too difficult, students are not capable of solving the problem.			
Encouraging reasoning	• Reasoning requires students to pull together patterns, connections, and understandings about the rules of mathematics, and then apply their insight into finding a solution to a difficult, challenging problem.			

This form is for leaders only.

We do not recommend this form be distributed.

While it is nonevaluative since no performance criteria are identified, the names make the list confidential.

Source: ©LCM 2012, Hull, Harbin Miles, and Balka, MathLeadership.com

actions (Figures 9.2 and 11.2) for student and teacher action relationship. The categories for this form are essentially the same as teacher actions and directly relate to the Matrix.

The only purpose for the Teacher Implementation Progression form is to assist leaders responsible for providing professional training to know what support is needed and who needs it. For the most part, direct classroom assistance is the best. Of course, this process works best when teachers work with leaders to realistically self-assess their progress and training needs.

As leaders such as coordinators, specialists, and coaches work with teachers on a regular routine, the information they must recall becomes staggering. Classroom visits help these leaders see general trends and indicate system progress. Yet individual teacher needs become blurred. Who is using which strategy? What level of proficiency have they achieved? What assistance do they need? How are students responding? What strategy do individual teachers need to add next, and who is ready to add another strategy?

Once leaders have a positive working relationship with teachers in using the Classroom Visit Tally–Teachers form, they may wish to incorporate the Teacher Implementation Progression (TIP). The actions on the TIP form are highly related to the Visit Tally and are directly related to the Matrix. Leaders want to sit quietly in a private location and reflect on their work with the teachers. With the form as a guide, leaders mark in pencil their first thoughts about where individual teachers are perhaps located.

At the end of this activity, leaders will most likely have some questions regarding where certain teachers are operating. As a result, leaders need to verify their selections. Because there are high correlations among the Matrix, Classroom Visit Tally–Teachers form, and TIP form, leaders may encourage teachers to discuss where they think they are in regard to the Matrix or Visit Tally at the next professional conversation related to classroom visit data.

For instance, at the next meeting leaders may ask teachers to think about the data indicating which strategies are being used with what frequency. Next, leaders discuss the three degrees of implementation with the teachers (experimenting, using, and integrating). Finally, leaders ask teachers to think about their classrooms and where they think they, and their students, fit. Teachers share if they are willing.

As a reminder, the criteria are provided, but the form with individual names is not. Teachers who are willing to share do so. Leaders have follow-up conversations with individual teachers concerning their needs. Leaders and teachers reach agreement on next steps.

Having Productive Conversations

1. What concerns do you have as teachers with the data recommended in this chapter?
2. What concerns as leaders do you have with the data recommended in this chapter?
3. How can these concerns be appropriately addressed, yet ensure progress is being made toward implementing the Practices?
4. How are leaders currently gathering data on instructional strategies?
5. How are teachers improving their Practices?
6. What are the current communication channels between leaders and teachers concerning instruction, the Practices, and student achievement?
7. When the time is right, how should elements in this chapter be initiated?

13

Solution Step 5

Maintaining Progress Toward Rigor

Instituting and achieving rigor both consist of complex and multidimensional factors. Rigor is a goal that emerges as the previous four steps (Chapters 9 to 12) are carefully and strategically implemented over time. While rigor has remained elusive in the past, the Common Core content and Standards for Mathematical Practice provide a rich and nurturing environment for rigor to develop. Certainly, rigor is deeper mathematical content understanding. But just attempting to teach content more deeply without altering the classroom and school learning culture and climate will prove to be unsuccessful. According to the National Research Council (NRC, 2012)

> We define "deeper learning" as the process through which an individual becomes capable of taking what was learned in one situation and applying it to new situations (i.e., transfer). Through deeper learning (which often involves shared learning and interactions with others in a community), the individual develops expertise in a particular domain of knowledge and/or performance. The product of deeper learning is transferable knowledge, including content knowledge in a domain and knowledge of how, why, and when to apply this knowledge to answer questions and solve problems. (pp. 5–6)

Since mathematical rigor is the desired goal, then leaders and teachers must monitor progress toward attaining rigor.

Background

In Chapter 2, "Defining and Instituting Rigor," mathematical rigor was clearly explained. Rigor was also directly tied to the Standards for Mathematical Practice and the Proficiency Matrix. In Chapters 9 through 12, four iterative steps were identified for leaders and teachers to take to effectively implement the Practices:

1. Monitoring Student Actions Related to the Practices
2. Using Classroom Visit Data—Assessment of Student Actions

3. Monitoring Teacher Actions Related to the Practices
4. Gathering and Using Additional Data

The ultimate outcome of a successful implementation of the steps over time is mathematical rigor. Incremental adoption of selected strategies, monitoring implementation of the strategies, and pertinent feedback are required procedures. While adoption and implementation are vital, they are not the desired outcome. Students must be the beneficiaries of the efforts to add strategies and change both classroom instruction and climate. For this reason, leaders and teachers, guided by the leadership teams, must routinely assess progress on instituting mathematical rigor.

Research offers important ideas that need to be more globally considered. Some research is specific but not necessarily helpful unless considering the implications of the research. For instance, the NRC (2000) states that learning means that the learner can transfer the learning to new or novel situations and that learning requires an effective and efficient retrieval system. These findings are rather difficult to spot. The Standards for Mathematical Practice offer statements such as the following:

▶ Make sense of problems and persevere in solving them.
▶ Reason abstractly and quantitatively.
▶ Construct viable arguments.

While more easily identified than perhaps transfer of knowledge, these Practice statements (and the other ones) are still difficult to spot. However, the implications of the Practices and of NRC's research are easily recognized in classrooms. Students must be working in groups of two to four on challenging problems, and they must be provided multiple opportunities to share and discuss their solutions. As we have indicated several times, students must be engaged in activities that require thinking, reasoning, and discussing.

Relating Mathematical Rigor and the Practices

Mathematical rigor is deep reasoning, thinking, and understanding related to mathematical content. As students engage in the Standards for Mathematical Practice and increase their level of proficiency, they are required to think, reason, and understand mathematics. For this reason, proficiency in the Practices, as exhibited by the three levels in our Proficiency Matrix, is highly correlated to mathematical rigor.

While mathematical rigor is directly related to deep understanding of content, rigor is not attainable by directly teaching only content. There are other factors required for rigor to both exist and thrive inside classrooms. Furthermore, rigor attainment is difficult to directly assess, so indicators that promote and sustain rigor must be identified, implemented, monitored, and assessed.

When reviewing research on factors that promote rigor, the same ones that promote deep reasoning, thinking, and understanding, we find they can be grouped, unsurprisingly, beneath four categories: content, instruction, assessment, and climate. The combination of and the interaction among these four categories support mathematical rigor. While the four categories are not unexpected, the idea that they are developmental, sequential, and iterative might be. As an example, positive classroom or school climate emerges as content, instruction, and assessment shift toward more challenging tasks, engaging instructional strategies, and higher cognitive demand assessments. The shift toward rigor is also iterative. In other words, a prior category need not be "perfect" before moving forward. With every iteration of the entire process, each category continually improves.

We previously discussed the potential impact of technology related to assessments. With assessments increasing in rigor, leaders and teachers may believe that assessment is the first category to address. While increasing the rigor of assessments can supply a sense of urgency, the category of assessment must be addressed through content and instruction. If students are not taught the appropriate content through instructional strategies that encourage thinking, reasoning, and understanding, assessment results are essentially invalid—violating the age-old rule of a written, taught, and tested curriculum (English, 2000). The data generated from the assessments do not reflect what the students were taught and are not, therefore, a fair reflection of what they know.

Inferences From the Standards for Mathematical Practice

By spending time with the Practices and the Proficiency Matrix, greater depth of understanding concerning the Practices emerges and how they are used to teach the Common Core content. While not explicitly stated, there are inferences that can be easily drawn. These inferences form the foundation for the categories and indicators in the Rigor Analysis Form located in Figure 13.1.

Inferences on Content

When reading the Practices, it is obvious that content is not part of the list. The Practices are not content, but rather a guide for how the content is taught. The inference is a rather easy connection to make. The content for the Practices is the Common Core mathematics content. When analyzing for rigor, clearly content is a viable category.

Content	1 Not Evident Minimal, Skill-Based	2 Skill/ Procedure Focus	3 Moderate Concept Inclusion	4 Concept Development Evident	5 High-Quality Content, Materials Incorporated
Grade-Level Appropriate					
Related to Conceptual					
Meaningful, Logical Flow Within Lesson					
Connected to Prior/Future Component					

Inferences on Instruction

To use the Practices for students to learn the Common Core content, the wording of the Practices must be studied as suggested in Chapter 4. The Practices state that students are to make sense of problems, reason, construct viable arguments, critique, model, and use tools correctly and appropriately. To achieve these demands, students must be provided challenging problems, time to think about and solve problems, time to discuss and share their thinking and understanding, and use models and tools to support their thinking, reasoning, and justifications. For students to be engaged in such activities, teachers must provide lessons that support these types of activities. There is an ongoing dynamic in the classroom of discourse between teachers and students and between students and students. Instruction, therefore, is a valid category for the Rigor Analysis Form.

Instruction	1 Not Evident Minimal, Skill-Based	2 Skill/ Procedure Focus	3 Progress Toward Indicators Apparent	4 Students' Participatory Routines Apparent	5 High-Level Student Engagement
Engaging					
Challenging Problems					
Interactive Discourse/ Dialogue					
Knowledge Transfer					
Open-Ended/Multiple Solution Paths					

Inferences on Assessment

As indicated previously, classrooms that use the Practices to help students learn mathematics must be interactive and have highly engaged students. If students are required to think and reason while solving challenging problems, then assessments (to maintain the notion of a written, taught, and tested curriculum) must reflect problem solving, thinking, and reasoning. Assessments must be varied and draw from many sources. Assessments, then, are a valid category for the Rigor Analysis Form.

Assessment	1 Not Evident, Did Not Occur	2 Skill, Procedure Focus	3 Problem Application	4 Performance Task	5 Authentic Real-life Problem
Content Appropriate					
Incorporates Thinking and Reasoning					
Formative Assessment Within Classroom Period					
Summative Assessment					

Inferences on Climate

Content is more rigorous, classrooms are more interactive, and assessments are more multidimensional. Yet all is for naught if students are not willing participants. For students to willingly participate in these dramatic shifts, classroom climate must shift accordingly. Students must feel valued, respected, and safe to share their thinking, understandings, and answers. Learning becomes community, not solitary. Teachers have to step off the stage and allow students to be the focus. There is a giant shift in beliefs that support the few in being successful to a belief that all are successful. Climate must be a category for the Rigor Analysis Form.

Climate	1 Teacher Dominated, Docile Students	2 Teacher Controlled, Top Tier Students Involved	3 Participation Promoted, Students Engaged	4 Encouraging Pushing Thinking for All Students	5 Student Focused, Teacher Facilitated
Everyone Involved					
Enthusiasm Apparent					
Positive Tone, Supportive					
Self-Assurance, Effective Effort					
Community of Learners					

Rigor as an Outcome

Rigor is the result of significantly shifting instructional "inputs" with an understanding and clear image of the desired "outcomes." Leadership teams and leaders focus on the path while periodically checking the horizon to ensure progress is being made toward the desired goal. Each educational position has its role, but the roles work in conjunction to achieve the common goal.

Categories

In monitoring and providing feedback on progress toward mathematical rigor, four categories can be identified:

1. Content
2. Instruction
3. Assessment
4. Climate

Each category is an important element of rigor. Each category also has specific indicators that can be used to guide progress toward mathematical rigor. The indicators are based on research and the Practices.

Content

The mathematics content being taught in the classroom must be grade level or course appropriate. The lesson content must be a meaningful component that is related to a concept students have been studying. The Common Core provides guidance for the domains and big ideas. The content should also be linked to prior and future learning.

Instruction

During classroom instruction, students need to routinely collaborate with their classmates. They should work on challenging problems that, when possible, are realistic in nature and of interest to the students. The problems should be open-ended with multiple solution paths. All students should be actively engaged frequently during a lesson. Discourse and discussion should be elevated beyond fact recall.

Assessment

One of the most important, but rarely used, forms of assessment is ongoing classroom formative assessment. While any assessment that is used to inform instruction is useful, timing for intervention is critical. Feedback that occurs within the classroom is extremely beneficial to students. Understanding mathematics is a complex undertaking, and students need constant monitoring to both check their current level of understanding and immediately correct any misunderstandings.

Climate

More challenging content taught in active, engaging classrooms does not occur without a supporting climate. Students must feel they are an important part of a learning community that respects everyone. In a positive climate, everyone participates. There is enthusiasm evident, a positive and supportive tone, and a sense of confidence.

The categories, with specific indicators, are organized to assist leadership teams and leaders in their efforts to achieve mathematical rigor.

Rigor Analysis Form

To assist leaders and teachers in reaching the goal of mathematical rigor, a Rigor Analysis Form (Figure 13.1) is provided. The form is developed around the four categories, and it provides teachers and leaders an opportunity to step back and carefully consider whether the elements needed for rigor are in evidence and to what degree they are apparent. In this way, leaders and teachers evaluate their efforts at implementing strategies and decide what next steps should occur. The form may be used to assess curriculum documents, resource materials, classroom instruction, or mathematics programs.

Explanation

Mathematical rigor is depth of thinking, reasoning, and understanding. Educators quickly discover that proving students have attained rigor is elusive, yet there are distinct indicators. As a result, achieving mathematical rigor becomes a process of ensuring the indicators for identified research-based categories are successfully instituted. As we noted earlier, rigor appears in four key categories: content, instruction, assessment, and climate. Categories are sequenced by developmental order of importance to progress, and the process is iterative. The indicators within each category are also sequenced. The Rigor Analysis Form is used to guide decisions and chart progress toward a viable, challenging mathematics program.

Directions

By recording ratios in the boxes, an analysis may be completed for a school, department, or grade level. For instance, if there are 15 teachers in a school, then the scores for each cell would have a denominator of 15. For individuals, materials, programs, and the like, using checkmarks in the cells may suffice. Ratings are based on a minimum of three visits or reviewers.

Figure 13.1 Rigor Analysis Form

Content	1 Not Evident Minimal, Skill- Based	2 Skill/Procedure Focus	3 Moderate Concept Inclusion	4 Concept Development Evident	5 High-Quality Content, Materials Incorporated
Grade-Level Appropriate					
Related to Conceptual (Big Ideas)					
Meaningful, Logical Flow Within Lesson					
Connected to Prior/ Future/ Component					
Instruction	1 Not Evident Minimal, Skill- Based	2 Skill/Procedure Focus	3 Progress Toward Indicators Apparent	4 Students Participatory, Routines Apparent	5 High-Level Student Engagement
Engaging					
Challenging Problems					
Interactive Discourse					
Knowledge Transfer					
Open-Ended/Multiple Solution Paths					

Assessment	1 Not Evident, Did Not Occur	2 Skill, Procedure Focus	3 Problem Application	4 Performance Task	5 Authentic Real-life Problem
Content Appropriate					
Incorporates Thinking and Reasoning					
Formative Assessment Within Classroom Period					
Data Informs Instruction					
Summative Assessment					

Climate	1 Teacher Dominated, Docile Students	2 Teacher Controlled, Top-Tier Students Involved	3 Participation Promoted, Students Engaged	4 Encouraging Pushing Thinking for All Students	5 Student Focused, Teacher Facilitated
Everyone Involved					
Enthusiasm Apparent					
Positive Tone, Supportive					
Self-Assurance, Effective Effort					
Community of Learners					

Source: © LCM 2013, Hull, Balka, and Harbin Miles, Mathleadership.com

Guiding the Work

After the Rigor Analysis Form is complete, leadership teams need to assess the current operating conditions as well as the area of immediate focus for improvement. The Rigor Analysis Focus Guide, Figure 13.2, is provided to assist in the process. Leadership teams and leaders should start at the top of the form and work their way down until they detect an area needing improvement.

Leaders, teachers, and leadership teams have access to three very important documents that are used to promote mathematical rigor— the Proficiency Matrix, Rigor Analysis Form, and the Rigor Analysis Focus Guide. These three documents build a common understanding of mathematical rigor, help establish clear expectations, and emphasize the actions and conditions that support rigor. These documents, plus the definitions of mathematical rigor from Chapter 2, clearly delineate and detail what mathematical rigor is and what it looks like when present.

Having Productive Conversations

1. What is the relationship between "deeper learning" and mathematical rigor?
2. How do the four areas impact rigor?
3. How can you use the Rigor Analysis Form?
4. How can you use the Rigor Analysis Focus Guide?

Figure 13.2 Rigor Analysis Focus Guide

What is being analyzed (school, grade level, department, classroom materials, etc.) ?		
Has content been rated as "Moderate Concept Inclusion" or higher?		
Yes		**No**
Do you have more than one independent rating source?		What are the content issues?
Yes Proceed	No Verify	How are these issues resolved?
Has instruction been rated as "Progress Toward Indicators Apparent" or higher?		
Yes		**No**
Do you have more than one independent rating source?		What are the content issues?
Yes Proceed	No Verify	How are these issues resolved?

(Continued)

Figure 13.2 (Continued)

Has assessment been rated as "Problem Application" or higher?	
Yes	**No**
Do you have more than one independent rating source?	What are the content issues?

Yes Proceed	No Verify	How are these issues resolved?

Has climate been rated as "Participation Promoted, Students Engaged" or higher?	
Yes	**No**
Do you have more than one independent rating source?	What are the content issues?

Yes Proceed	No Verify	How are these issues resolved?

Rigor Analysis
What category is the one needing immediate attention? Why?
What actions need to be taken, by whom, and by when? _____Approved
Will these actions have a positive effect on the next category?
What is the date of the next analysis?
Who is responsible for meeting the deadline? _____Approved

Source: © LCM 2013, Hull, Balka, and Harbin Miles, Mathleadership.com.

PART IV

Inputs and Outcomes

Readers have extensively studied mathematical rigor, how to attain rigor, possible problems, and workable solutions. In the following sections, we highlight changing the system. Schools are highly developed operating systems with standardized structures and norms. For real change to occur, the system (operating orders and norms) must change. In this respect, schools may be viewed as functions. Functions have inputs and outcomes. Leaders and leadership teams determine the inputs and then monitor outcomes.

The focus of this book is on implementing the Standards for Mathematical Practice, with the goal of instituting mathematical rigor. To successfully support the implementation of the Practices and achieve the goal, leaders and leadership teams are provided suggestions for specific actions that need to happen. Two of the suggestions for action are inputs, and two of them are outcomes. The inputs are curriculum and classrooms. The outcomes are communication and culture. As leaders and teachers strive to change individually, they must also focus on changing systemically.

14

Teaching for Rigor

Teachers are the essence of change. If mathematics leaders and school leaders do not actively and directly work to support teacher improvement efforts, change will not occur and student learning will not improve. Furthermore, if teachers are not attending to what students are actually doing, saying, thinking, and demonstrating, then the Common Core Standards will fail and the hope for mathematical rigor will dissolve.

Inputs

Curriculum

When working on curriculum, teachers need to guarantee that the grade-level or course material they are using matches what the Common Core outlines. As teachers plan lessons, they need to carefully study what mathematical knowledge students must obtain and then plan lessons that have a high degree of probability that students will successfully learn the content. Once planned, teachers strive to present the lesson as written and make only major deviations when absolutely necessary.

Teachers also work to promote connections among the content presented. Without taking care, students fail to relate one day's lesson to the next. Teachers must insist students have multiple opportunities to connect their learning and understanding to prior knowledge. Connections are more easily made when students have adequate time to make sense of the content and are able to work more deeply with it.

Lessons should take an integrated approach whenever possible. Skills should be taught in context. If a skill needs to be taught one or two days before it is used, students should clearly be informed concerning how the skill is important, and why it will be needed for the upcoming lesson.

Classroom

Many instructional approaches have been addressed. Within their classrooms, teachers routinely monitor student understanding by employing classroom formative assessment techniques. One required technique is for students to work collaboratively. Teachers facilitate student conversations so that all students have multiple opportunities to express their thoughts. These conversations, if students are explaining their thinking, must be fostered through engaging and challenging problems.

Teachers work with their students to teach them to value other people's ideas and thoughts. Also, students need help in actually learning to think and reason, capturing their thinking and reasoning processes, and then verbalizing the processes in a clear, articulate manner. Classrooms must encourage and respect thinking.

Outcomes

Communication

Communication, as used here, is different from classroom discourse. Communication is the flow of knowledge and learning that takes place within the school. It is how people learn what they are expected to know and do. A critical part to initial communication is related to the school vision and mission. If schools are going to take time to formulate visions and missions, then the documents containing them should be worth the time and effort to read. A vision should set the tone for the school, and teachers should know and communicate that vision during meetings and to the students. The statement is not advocating that visions be memorized and repeated but that the vision is a powerful statement driving beliefs and actions. If a vision statement promotes the idea that all students can and will learn, then this statement needs to be lived out every day, in every classroom. Related to this idea is one of encouraging and promoting high expectations for all students learning. How this message is conveyed is a matter of communication. High expectations are set during all communications between teachers, leaders, and students. Having a common, clearly understood vision instills positive interactions and builds rapport. Rapport improves communication channels and increases productivity in planning meetings.

As leadership teams work to support change initiatives related to implementing the Standards for Mathematical Practice and CCSS content, communication improves. As communication improves, and as content and classroom actions improve, the system starts readjusting. If communication is not more positive, then the actions in the classrooms are most likely not working as desired.

Culture

As actions begin, leadership teams need to carefully monitor any shifts in school and classroom learning culture. While culture does take time to shift, failure to detect any progress is cause for concern. Leadership teams and leaders are monitoring students' actions within classrooms. If students are actively engaging in mathematics lessons and collaborating in small groups, then climate will shift.

Indicators for cultural shifts for teachers include evidence that all students are engaged in classroom activities. Teachers are encouraging and supportive of all students. Students and teachers demonstrate a positive attitude and show respect for ideas and the individuals expressing the ideas. There is also increased evidence that teachers and students are more willing to take risks and to try new approaches.

There is a noticeable shift in effort. Students and teachers express a sense of "growth mindset" (Dweck, 2006). In other words, learning is attributed to effort, not genetic inheritance. Students often demonstrate a "we can" attitude and enjoy taking on challenging problems. Finally, there is a constant push to continually improve.

Teaching for Progress in Rigor

Teachers are invaluable assets and must be treated with the respect they are due. Change is difficult, and people change for different reasons and at different rates. Because a teacher does not rush forward to embrace a change is not an indication of some serious problem, but actually the normal change process. With support and training, teachers will make necessary shifts in content and classrooms. As these shifts occur, additional, corresponding shifts occur in communications and climate, and rigor emerges. Teachers must realize that rigor is an outcome of a myriad of factors. As the factors are put into place, and monitored for success, teachers will note a definite rise in mathematical rigor from their students.

Having Productive Conversations

1. How do you think teachers can best monitor curriculum?
2. How do you think teachers can best monitor classrooms?
3. How do you think teachers can best monitor communication?
4. How do you think teachers can best monitor culture?

15

Coaching for Rigor

As pointed out within this book, the term "leaders" is multidimensional. One of the roles of formal leaders is to build leadership capacity among students and teachers. In this way, formal and informal leaders are empowered to sustain change and continually improve. Mathematics leaders need to maintain the improvement focus on individuals and systems. Mathematics leaders, such as coaches, supervisors, and specialists, have distinct roles and responsibilities. The roles and responsibilities of mathematics leaders are vital to change efforts since they work directly in classrooms. Since they play such a vital role, there is a real need for everyone to clearly understand the duties and support the leaders' efforts in focusing on realizing mathematical rigor in classrooms.

Coaches, supervisors, and specialists are in a unique position to support teachers during the change required by the Common Core Standards. These mathematics leaders can work side by side with teachers in classrooms. With a keen knowledge of mathematics content and pedagogy, these leaders fill an invaluable niche. By regularly being in classrooms, they have firsthand knowledge of shifts that are occurring with teachers and students, shifts that promote rigor.

Inputs

Curriculum

Mathematics leaders need to be highly involved in planning lessons with teachers and coteach as often as possible. In this way, mathematics leaders are better prepared to assure the appropriate content is indeed being taught and at a high level. Mathematics leaders have more time to study the content and reflect on instructional practices. This information can be carefully infused into both planning and presenting.

By constantly working with teachers and studying mathematics curriculum documents, mathematics leaders are able to assist teachers in being more intentional about learning progressions and concept development. They are also able to suggest an even balance in lessons that promote concept development and those that promote skill development.

As collaboration improves between teachers, mathematics leaders are in a position to capture the learning. With this information, they can assist in designing and distributing model lessons where rigor is a focus.

Classrooms

Mathematics leaders, who are successful in their position, regularly enter classrooms in a variety of ways and for a variety of reasons. These reasons are all directly related to improving student learning. Mathematics leaders need to visit classrooms to collect data on strategy usage and student actions.

When possible, and at least weekly, mathematics leaders need to coteach with teachers. With a new assessment demands to implement rigor, this collaboration is essential. In coteaching, the leader and the teacher share responsibilities for instruction. Coteaching is significantly different from demonstration teaching. During coteaching, neither the teacher nor the mathematics leader are ever sitting and observing as with demonstration teaching. Both individuals are always moving about the classroom, providing help as needed. During coteaching, teachers and mathematics leaders assume primary responsibility for different sections of the lesson, but collaboration is the desired state.

Mathematics leaders need to regularly be in classrooms assisting students. In these situations, a mathematics leader serves as another pair of eyes, ears, and hands. By working with students, mathematics leaders are better able to understand student difficulties and misconceptions. This information readily informs instructional practices.

Mathematics leaders also enter classrooms to observe. In most cases, this is not an evaluative observation. Mathematics leaders are generally not assigned to a role of formal evaluator, and they must be careful to not cross the line. Teachers and mathematics leaders need to have a certain degree of rapport and trust before observations are beneficial. Nonetheless, mathematics leaders need to gather information to inform them in what actions to take to support teachers as they undergo change.

Outcomes

Communication

Mathematics leaders are highly involved in supporting communication between teachers, within the school, and between teachers and school leaders. Mathematics leaders, as good communicators, must be great listeners. Mathematics leaders should do far more listening than speaking. Valuable information is obtained by demonstrating effective listening skills.

The attitude and outlook of a mathematics leader is apparent in every interaction. Mathematics leaders must constantly encourage and support teachers. Without appearing naïve or uninformed, mathematics leaders need to always remain positive, even in the face of challenges and difficulties.

In respect to the vision and expectations, mathematics leaders need to ensure that the message is always in the forefront. Student learning, success, and achievement are at the center of every meeting. Not only do mathematics leaders promote the message, they also carefully observe to see if the desired actions are having the desired effects.

Culture

Mathematics leaders can sense whether the classrooms and schools are becoming student oriented and if a learning community atmosphere is emerging. We have noted that rigor in mathematics requires active student engagement, communication, and collaboration. Leaders need to monitor these changes. Not only do mathematics leaders assess the changes, they actively support the changes. Mathematics leaders encourage the use of new instructional techniques. As the techniques are first being used, mathematics leaders are encouraging and supporting. They help analyze and reflect on the results, and offer suggestions for refining the technique.

In this way, mathematics leaders maintain a positive attitude, build collegiality, and promote continual professional learning. In this respect, mathematics leaders are not passive bystanders, but active participants. Mathematics leaders must be continual learners too.

Of course, not every effort will be an immediate success. In some cases, a technique may never work as intended. Mathematics leaders encourage the process of learning from both failures and successes. They demonstrate and participate in a "we can" approach.

Coaching for Progress in Rigor

The job outcomes of a mathematics leader are immeasurable. Mathematics leaders have opportunities to promote change at the very source of the needed change—classrooms. Mathematics leaders are able to spend time working with teachers, and teachers make the difference for students. Mathematics coaches are in a unique position to monitor and document change over time. Coaches can carefully analyze the instructional shifts that are occurring and identify the resulting growth of rigor.

Having Productive Conversations

1. How do you think mathematics leaders can best monitor curriculum?
2. How do you think mathematics leaders can best monitor classrooms?
3. How do you think mathematics leaders can best monitor communication?
4. How do you think mathematics leaders can best monitor culture?

16

Leading for Rigor

Leaders must garner support from individuals responsible for implementing the change—teachers. Leaders have the ability to make things happen. They work to arrange meetings and focus conversations. Leaders are major players in establishing a school's vision and mission. They have access to various types of student performance data, as well as information from their state concerning assessment.

While leaders have access and can open doors, they need teachers to actually make a difference for students. Furthermore, as we have stated, teachers need support from both leaders and other teachers to make significant change. As a result, leaders must empower teachers. Collaboration across all levels of the school system is critical.

Inputs

Curriculum

Leading requires collaboration. School leaders must have established processes to regularly work with both the mathematics leader and the teachers or teacher representatives. School leaders are not, and cannot be, experts in every content subject area. Leaders can, however, work directly with the individuals who do have content expertise.

School leaders are responsible for ensuring curriculum documents, materials, and resources are available. They work directly with mathematics leaders and teacher leaders to guarantee the documents and materials are being used as designed.

To guarantee the content materials are used appropriately, school leaders are responsible for establishing time for collaboration and planning. Moreover, school leaders ensure that collaborative decisions are based on up-to-date data. School leaders must routinely analyze performance data and share the data with all responsible parties.

Classrooms

While school leaders may not be content experts in mathematics, there are many actions leaders can take to ensure change is occurring. As often as possible, school leaders need to visit classrooms. These visits should purposefully focus on the current change initiative under way. If a strategy such as think, pair-share has been adopted, then leaders need to check for evidence that the strategy is being regularly used. School leaders are also required to conduct formal classroom observations. The important note here is for leaders to ensure that formal observations are not undermining change initiatives.

To sustain change, there needs to be evidence of success. In other words, teachers want to know that their efforts are valued and that their efforts are helping students learn. For this reason, school leaders need to seek and maintain evidence of progress.

Outcomes

Communication

As with so many things contained within a school, school leaders are the gatekeepers. Leaders are often unaware of all the messages they send to teachers. If teachers are unsure of what to do, they fall back to routines and previous actions. These routines and previous actions may have emerged by default. The leader has not clearly indicated what should take place, so everyone infers what the leader wants. Rumor and inadequate communication almost always support the current system of operating.

In times of change, a leader must supply clear, concise information concerning expectations. These expectations relate to the vision and information gained from leadership teams. Consistency in the message is an absolute must, and the only way to achieve consistency is through regular communication that repeats the same information. To this end, leaders need open channels of communication.

For effective communication to occur, leaders must be excellent listeners. Without carefully listening to what is being said, leaders have no way of evaluating the clarity of their message. Leaders should recall that body language frequently overrules language. Also, what is *not* said may be as powerful as what is said, and actions trump words. To achieve a clear, concise message, leaders need to focus on students' success. Leaders ask what is best for students, why it is best for students, and what is the evidence that indicates the decision is best.

Culture

In the previous section on communication, leaders learned about their role as gatekeepers. In the area of school climate, school leaders are the primary source for determining what the climate is and will become. The decisions leaders make and the actions leaders take set the stage for the necessary cultural shifts.

If classrooms are intended to be learning communities, then the school leader must operate the school as a learning community. If teachers and students demonstrate positive attitudes toward learning, leaders must demonstrate positive attitudes toward learning. By functioning in leadership teams, leaders can easily demonstrate shared responsibility, respect, appreciation, and a positive attitude.

When teachers and leaders work together in teams to promote student success, the atmosphere and climate of the meetings move into classrooms. The message is simple—*we can work together to overcome any difficulty*. As the teaming process is successfully implemented and classroom actions are shifted, a leader is able to recognize the success of the teachers' efforts.

Leading for Progress in Rigor

School leadership is changing. The demands on school leaders are immense. Five or more decades ago, school leaders were managers. They "managed" the facilities, the students, and the finances. School leaders managed, and teachers taught. Each role had clearly defined responsibilities that rarely intersected other than student discipline.

While school leaders still "manage" their facilities, they are now responsible for "leading" their staff. These responsibilities have been emerging and clarifying within the last decades. While most of the Common Core emphasis appears to currently focus on teachers and teaching, the leadership role is actually far more significant. The leadership responsibilities are looming right on the horizon and will draw national attention soon. After all, the Common Core Content and Practices will not be successfully implemented without effective school leadership.

One message is coming through loud and clear for teachers and school leaders—business as usual is no longer going to work. The shift is not going to be easy, but it can happen. School systems, and the underlying assumptions, are currently in balance. The culture and climate of schools and districts are so embedded that often the conditions are no longer even recognized. A "manager" is unable to successfully create and sustain the conditions for change. Leadership is the only option.

Leaders must carefully balance implementation of details while charting progress toward identified goals. Leaders need to develop a clear mental image of what rigorous mathematics classrooms look like, then guarantee that steady, reasonable progress is being made toward achieving the mental image—not in just one or two classrooms, but in each and every classroom.

Having Productive Conversations

1. How do you think school leaders can best monitor curriculum?
2. How do you think school leaders can best monitor classrooms?
3. How do you think school leaders can best monitor communication?
4. How do you think school leaders can best monitor culture?

PART V

Momentum

Sustaining momentum is necessary for success. This momentum heavily relies on the continued actions of formal leaders. If leaders withdraw their support, become sidetracked with other issues, or merely quit paying attention, the momentum will slow to a halt as will implementation. There are specific actions mathematics teachers, mathematics leaders, and school leaders need to take.

Momentum also consists of constantly monitoring progress on the degree of implementation. Real progress, as indicated earlier in the book, means actions change, but the system also changes. Leaders and leadership teams must monitor the "pieces" to ensure student actions, teacher actions, and leaders' actions are occurring as desired. Leaders and leadership teams must monitor that the enacted pieces are creating the desired outcomes and ensuring the goal of mathematical rigor.

17

Linking Responsibilities— Assessing Progress

Much information has been provided in the book regarding rigor and the Standards for Mathematical Practice. There are many tools to assist with integrating and sustaining change related to the Common Core. This chapter provides one final tool for leaders to use as they continue to make progress. The tool is designed to demonstrate a relationship among the responsibilities of students, teachers, mathematics leaders, and school leaders.

The tool is called Linking Responsibilities Actions/Indicators, and it is a chart that pulls key ideas from the book into a single document. By using the chart as a guide, and routinely assessing the current actions being undertaken, leaders can focus on specific areas for improvement. The tool assists leaders in tracking details and charting progress toward effective implementation of the Practices and, ultimately, mathematical rigor.

With teacher assistance, leaders have taken steps using classroom visits to regularly monitor shifts in classroom instruction to support teachers during the transitional change time.

Teachers and leaders are focusing on student actions that directly relate to the Standards for Mathematical Practice. They have also started conversations that relate specific teacher actions to the student actions. These two areas are the first two steps in the Change Chain Reaction.

Students must be engaged in specific, research-informed actions to learn mathematics.

For students to be engaged in these actions, teachers must appropriately use specific strategies and techniques.

For teachers to know and use these strategies and techniques effectively, leaders must engage in specific, consistent actions.

By thinking carefully about the Change Chain Reaction, teachers and leaders realize the difficulty in actually separating student actions, teacher actions, and leader actions. This idea is emphasized in the discussion concerning system operations. All three levels of action are a self-fulfilling loop. Even though the actions are highly interrelated, there is still a need to highlight each step in the Change Chain Reaction. With this thought in mind, we want to shift to Statement 3, the role of the leader before offering the Linking Responsibility Chart.

Professional Trust

Teachers must trust their leaders, and it is the leader's responsibility to ensure the teachers are provided with every justification and reason to do so. Leaders must treat teachers with respect, honor their opinions, listen carefully, and abide by a code of strict confidentiality. As a way to build and support trust, we encouraged the process of focusing first on students, promoting productive conversations, and monitoring strategy implementation rather than teacher evaluation. When the classroom visits shifted to teacher actions, we stated that teachers' names were not included on the form. The focus was on systemwide implementation of specific student and teacher actions. By focusing on the students, teachers worked among themselves, with leader support, to determine how to get students more actively engaged.

Professional Conversations

Throughout the change process, professional conversations have routinely occurred. These conversations are teacher to teacher and teachers to leader. The level of professional conversation should, by now, be greatly increased. The conversations focus on effective strategies, student learning, and the Practices. These conversations are centered on classroom visit data, the Matrix, visible thinking, and formative assessment. Additional data sources, if not already discussed, may be included. These data sources, both formal and informal, are compared to changes happening in classrooms.

Supporting Teacher Change

Chapter 10, "Solution Step 2: Using Classroom Visit Data—Assessment of Student Actions," contains information about how adults generally change behaviors. The process includes a willingness to adopt a strategy and a series of steps to actually perfect using a strategy. Failure to understand these two areas of change has perhaps been the greatest support for maintaining the current operating system. By ignoring what is known about change, efforts to improve have been little more than a speed bump to the system status quo. In spite of all the possible fanfare and hype, actual change procedures were never enacted. Thus, the operating system continued to operate as usual.

If the Common Core Standards for Mathematical Practice are to be realized, the typical pattern of offering change must actually be changed itself. This responsibility resides with leaders. Leaders must understand the change process. Understanding change includes knowing how to build communities of learners and leadership teams; collect, analyze, and use data; recognize the Practices in action; and support teachers' change efforts.

Working to Improve

Leaders and teachers need to work together to adopt and implement the Standards for Mathematical Practice. This process requires collegial relationships based on trust and honesty. Teachers must be honest concerning their progress, and leaders must be honest concerning their expectations. Mathematics leaders, the ones identified in this chapter, have various resources they can use to support teachers. The closer to actual classroom instruction the support is the more effective it will be.

Teachers need to recognize that leaders have to be present in classrooms on a regular basis. Leaders also need to work with students whenever possible. Leaders may coplan, coteach, visit, or observe (nonevaluative). Leaders may also assist teachers in analyzing various forms of data such as visit data, assessment data, or student work. Leaders also assist teachers in reflecting on their practice.

Coplanning and coteaching are excellent opportunities to implement new strategies or perfect existing ones. Leaders promote individual teacher growth. They may conduct training sessions, but they should also promote teachers leading training sessions. With leader organization, teachers can support their fellow teachers in coplanning, coteaching, analyzing, and reflecting.

Documenting Change

If the desired actions are indeed occurring, and if monitoring results support that this is the case, then both inputs and outcomes are shifting. However, merely believing the changes are occurring is not sufficient. School leaders, mathematics leaders, teachers, and students must recognize that changes are happening.

In Chapters 14, 15, and 16, the ideas concerning inputs (content and classroom) and outcomes (communication and climate) were explained. These four areas—content, classroom, communication, climate—must be monitored by the four groups impacted—students, teachers, mathematics leaders, and school leaders. The Linking Responsibilities Chart (Figure 17.1) assists in organizing the data.

Teachers and leaders must reach consensus concerning the actions. Misunderstandings and misinterpretations are very likely at the initial phases of implementation. Furthermore, understandings may shift as greater expectations emerge. Early stages of encouraging thinking beneath teacher classroom input should not be at the same level of proficiency throughout implementing the Practices.

Using the Form

The form is nonevaluative and may be used by any educator in any position. The form works well in collaborative conversations concerning progress with representation from several, or all, of the four identified groups—students, teachers, mathematics leaders, and school leaders. The significance of the form is to ensure that each group is making progress within each area. Furthermore, the form highlights starting with inputs and then ensuring desired outcomes are emerging.

Conclusion

Mathematical rigor is definable, understandable, implementable, and achievable. Our students are well worth the energy and effort. Their futures are highly dependent on the decisions school leaders make. We believe the conclusion is obvious.

Conclusion: Mathematical rigor is attained by teaching the Common Core content using the Standards for Mathematical Practice to the depth indicated within the Proficiency Matrix.

There is no time like the present to plan for the future.

Figure 17.1 Linking Responsibilities Actions/Indicators

	Students	Teachers	Math Leaders	School Leaders
Content (input)	• Do grade-level appropriate work. • Understand connections. • Work at conceptual level. • Gain proficiency or skill.	• Provide grade-level appropriate lessons. • Stress instructional connections. • Provide depth in lessons. • Provide skills in context.	• Assist in planning and teaching appropriate lessons. • Regularly discuss progressions. • Regularly discuss development of concepts. • Monitor for balance in lessons for concepts and skills. • Design and model content in appropriate lessons.	• Collaborate with math leaders. • Collaborate with teachers. • Ensure content materials. • Provide collaborative time. • Analyze student data.
Classroom (input)	• Engage in active participation. • Work as learning community. • Openly discuss reasoning. • Participate in mathematical discourse.	• Monitor learning. • Provide collaborative learning opportunities. • Encourage thinking. • Facilitate conversations.	• Coteach with teachers. • Visit classrooms often. • Work with students. • Observe.	• Visit classrooms often. • Observe. • Chart progress.
Communication (outcomes)	• Express interest in math. • State they can learn math. • Articulate need for math. • Talk positively about math.	• Discuss school vision. • Promote high expectations. • Demonstrate rapport. • Participate in collaborative planning.	• Demonstrate listening. • Show positive encouragement. • Share information. • Speak encouragingly.	• Promote the vision. • Actively listen. • Communicate. • Promote goals for success.
Climate (outcomes)	• Take risks. • Help other students. • Have positive attitudes. • Be responsible for learning.	• Support all students. • Try new approaches. • Promote continuous improvement. • Have "can do" approach.	• Support change. • Encourage new techniques. • Have positive "we can" attitude. • Build collegiality.	• Share leadership. • Recognize success. • Promote "we can" attitude. • Continuously improve.

Having Productive Conversations

1. How can the Linking Responsibilities Chart best be used to promote professional learning?
2. In early stages of using the Linking Responsibilities Chart, which indicators should reasonably be checked?
3. How do deeper understandings of the Practices develop with use of the Linking Responsibilities Chart?
4. Thinking about change over time and realistic expectations concerning learning and integrating strategies, how would the Linking Responsibilities Chart data evolve with the school year?
5. How can teachers and leaders work together to ensure appropriate, yet realistic, progress?

Appendix A

Standards of Student Practice in Mathematics Proficiency Matrix

	Students:	(I) = Initial	(IN) = Intermediate	(A) = Advanced
1a	Make sense of problems.	Explain their thought processes in solving a problem one way. *(Pair-Share)*	Explain their thought processes in solving a problem and representing it in several ways. *(Questioning/Wait Time)*	Discuss, explain, and demonstrate solving a problem with multiple representations and in multiple ways. *(Grouping/Engaging)*
1b	Persevere in solving them.	Stay with a challenging problem for more than one attempt. *(Questioning/Wait Time)*	Try several approaches in finding a solution, and only seek hints if stuck. *(Grouping/Engaging)*	Struggle with various attempts over time, and learn from previous solution attempts. *(Allowing Struggle)*
2	Reason abstractly and quantitatively.	Reason with models or pictorial representations to solve problems. *(Grouping/Engaging)*	Translate situations into symbols for solving problems. *(Grouping/Engaging)*	Convert situations into symbols to appropriately solve problems as well as convert symbols into meaningful situations. *(Encouraging Reasoning)*
3a	Construct viable arguments.	Explain their thinking for the solution they found. *(Showing Thinking)*	Explain their own thinking and thinking of others with accurate vocabulary. *(Questioning/Wait Time)*	Justify and explain, with accurate language and vocabulary, why their solution is correct. *(Grouping/Engaging)*
3b	Critique the reasoning of others.	Understand and discuss other ideas and approaches. *(Pair-Share)*	Explain other students' solutions and identify strengths and weaknesses of the solutions. *(Questioning/Wait Time)*	Compare and contrast various solution strategies, and explain the reasoning of others. *(Grouping/Engaging)*

	Students:	(I) = Initial	(IN) = Intermediate	(A) = Advanced
4	Model with mathematics.	Use models to represent and solve a problem, and translate the solution into mathematical symbols. *(Grouping/Engaging)*	Use models and symbols to represent and solve a problem, and accurately explain the solution representation. *(Question/Prompt)*	Use a variety of models, symbolic representations, and technology tools to demonstrate a solution to a problem. *(Showing Thinking)*
5	Use appropriate tools strategically.	Use the appropriate tool to find a solution. *(Grouping/Engaging)*	Select from a variety of tools the ones that can be used to solve a problem, and explain their reasoning for the selection. *(Grouping/Engaging)*	Combine various tools, including technology, explore, and solve a problem as well as justify their tool selection and problem solution. *(Allowing Struggle)*
6	Attend to precision.	Communicate their reasoning and solution to others. *(Showing Thinking)*	Incorporate appropriate vocabulary and symbols in communicating their reasoning and solution to others. *(Allowing Struggle)*	Use appropriate symbols, vocabulary, and labeling to effectively communicate and exchange ideas. *(Encouraging Reasoning)*
7	Look for and make use of structure.	Look for structure within mathematics to help them solve problems efficiently (such as $2 \times 7 \times 5$ has the same value as $2 \times 5 \times 7$, so instead of multiplying 14×5, which is $[2 \times 7] \times 5$, the student can mentally calculate 10×7). *(Question/Prompt)*	Compose and decompose number situations and relationships through observed patterns in order to simplify solutions. *(Allowing Struggle)*	See complex and complicated mathematical expressions as component parts. *(Encouraging Reasoning)*
8	Look for and express regularity in repeated reasoning.	Look for obvious patterns, and use if/then reasoning strategies for obvious patterns. *(Grouping/Engaging)*	Find and explain subtle patterns. *(Allowing Struggle)*	Discover deep, underlying relationships (uncover a model or equation that unifies the various aspects of a problem such as discovering an underlying function). *(Encouraging Reasoning)*

Source: © LCM 2011, Hull, Balka, and Harbin Miles, Mathleadership.com

Appendix B

Instructional Implementation Sequence

Strategy	Description
Initiating think, pair-share	Pair-share, or think, pair-share, is a strategy easy to implement in any classroom at any grade level or subject. This strategy does not require any other change in pedagogy or materials. For pair-share, teachers merely ask a question or assign a problem and allow students to think and work with a partner for one to three minutes before requesting an answer to the question or problem. In think, pair-share students are given a brief period of time to think independently before working with a partner. While effective in results, this strategy is a significant first step in engaging all students in classroom instructional activities.
Showing thinking in classrooms	Teachers need to work toward higher degrees of student involvement in classroom activities. Once pair-share is incorporated into classroom routines, teachers need to incorporate additional strategies that promote "every pupil response" (EPR). EPR strategies include such responses as "thumbs up/thumbs down," or use of individual whiteboards for noting answers. Students are also pressed to be more aware of their thinking and express their thinking in more detail. Students are routinely asked to share their thinking in mathematics classrooms. However, what is routinely accepted as thinking is actually process description. Students merely provide the steps they used to solve the problem, not their reasoning and thinking about how they knew which processes to use. To reveal student thinking, more challenging, open-ended problems are needed.
Questioning and wait time	As thinking is increased in the mathematics classroom, better questioning and wait time are required. Teachers need to provide thought-provoking questions to students, and then allow the students time to think and work toward an answer.

Strategy	Description
	Empowerment Strategies
Grouping and engaging problems	The strategy of "grouping and engaging problems" is a significant shift in pedagogy and materials. Students are given challenging problems to work, and they are allowed to work on the problem in a group of two, three, or four. Challenging mathematics problems take time, effort, reasoning, and thinking to solve.
Using questions and prompts with groups	Once students are provided with opportunities to solve challenging problems in groups, teachers need to increase their ability to ask supporting questions that encourage students to continue working, provide hints or cues without giving students the answers, and ask probing questions to better assess student thinking and current understanding.
Allowing students to struggle	Students learn to persevere in solving challenging mathematics problems by being allowed to struggle with challenging problems. Students need to understand that mathematical problems do not usually have a quick, easy solution. Effective effort is a life-skill and should be learned interdependently and independently. Appropriate degree of difficulty is foremost on teachers' minds. If the problem is too easy, students do not need to struggle. If the problem is far too difficult, students are not capable of solving the problem. Teachers need to balance working in groups and working independently, and be able to quickly adjust grouping strategies as the need arises.
Encouraging reasoning	Students need to be encouraged to carefully think about mathematics and to understand their level of knowledge. They also need to be able to accurately communicate their thinking. Reasoning, in this context, is used to convey having students stretch their understanding and knowledge to solve challenging problems. Reasoning requires students to pull together patterns, connections, and understandings about the rules of mathematics, and then apply their insight into finding a solution to a difficult, challenging problem.

Source: ©LCM 2012, Hull, Harbin Miles, and Balka, MathLeadership.com

References

Balka, D., & Hull, T. (2011). *Visible thinking activities*. Rowley, MA: Didax.

Balka, D., Hull, T., & Harbin Miles, R. (2010). *A guide to mathematics leadership*. Thousand Oaks, CA: Corwin.

Boaler, J. (2008). *What's math got to do with it? Helping children learn to love their least favorite subject*. New York, NY: Viking.

Bransford, J., Brown, A., & Cocking, R. (2000). *How people learn: Brain, mind, experience, and school*. Washington, DC: National Academy Press.

Brookhart, S. (2008). Feedback that fits. *Educational Leadership, 65*(4).

Conference Board of the Mathematical Sciences. (2013). *Common core state standards for mathematics statement by presidents of CBMS member professional societies*. Washington, DC: CBMS.

Daggett, W. (2005). Achieving academic excellence through rigor and relevance. International Center for Leadership in Education. Retrieved from www.leadered.com/pdf/academic_excellence.pdf

Depth of Knowledge (DOK). *Levels for mathematics*. Retrieved from http://static.pdesas.org/content/documents/DOK_Math_levels.pdf

Dweck, C. (2006). *Mindset: The new psychology of success*. New York, NY: Random House.

English, F. (2000). *Deciding what to teach and test*. Thousand Oaks, CA: Corwin.

Gavin, G., & Moylan, K. (2012). 7 steps to high-end learning. *Teaching Children Mathematics, 19*(3), 184–192.

Hall, G., & Hord, S. (2001). *Implementing change: Patterns, principles, and potholes*. Needham Heights, MA: Allyn and Bacon.

Heritage, M. (2011). Formative assessment: An enabler of learning. *Better: Evidenced-based education*. Spring, 18.

Herman, J. L., & Linn, R. L. (2013). *On the road to assessing deeper learning: The status of Smarter Balanced and PARCC assessment consortia*. (CRESST Report 823). Los Angeles, CA: University of California, National Center for Research on Evaluation, Standards, and Student Testing (CRESST).

Hull, T., Balka, D., & Harbin Miles, R. (2012). *The common core state standards: Transforming practice through team leadership.* Thousand Oaks, CA: Corwin.

Hull, T., Harbin Miles, R., & Balka, D. (2011). Overcoming resistance to change: Why isn't it working? *Virginia Mathematics Teacher, 38*(1), 36–38.

Iowa Department of Education. (2005). *Improving rigor and relevance in the high school curriculum.* Des Moines, IA: State Department of Education.

LCM. www.mathleadership.com

Literacy and Numeracy Secretariat. (2008). *Differentiating mathematics instruction.* Special Edition 7. Ontario, Canada. Retrieved from, http://www.edu.gov.on.ca/eng/literacynumeracy/inspire/research/different_math.pdf

Marzano, R. (2003). *What works in schools: Translating research into action.* Alexandria, VA: Association for Supervision and Curriculum Development.

Murphy, J., Hallinger, P., & Heck, R. H. (2013). Leading via teacher evaluation: The case of the missing clothes? *Educational Researcher, 42*(6), 349–354.

National Council of Teachers of Mathematics. (1989). *Curriculum and evaluation standards for school mathematics.* Reston, VA: National Council of Teachers of Mathematics, Inc.

National Council of Teachers of Mathematics (2000). *Principles and standards for school mathematics.* Reston, VA: National Council of Teachers of Mathematics, Inc.

National Governors Association Center for Best Practices and Council of Chief State School Officers (2010). *Common core state standards – Mathematics.* Washington, DC: National Governors Association Center for Best Practices and Council of Chief State School Officers.

National Mathematics Advisory Panel. (2008). *Foundations for success: The final report of the national mathematics advisory panel.* Washington, DC: U.S. Department of Education.

National Research Council. (1999). *Improving student learning: A strategic plan for education research and its utilization.* Washington, DC: National Academy Press.

National Research Council. (2000). *How people learn: Brain, mind, experience, and school.* Washington, DC: National Academy Press.

National Research Council. (2001). *Adding it up: Helping children learn mathematics.* Washington, DC: National Academy Press.

National Research Council. (2004). *Engaging schools: Fostering high school students' motivation to learn.* Washington, DC: National Academy Press.

National Research Council. (2005). *How students learn: History, mathematics, and science in the classroom.* Washington, DC: National Academy Press.

National Research Council. (2011). *Successful K–12 STEM education: Identifying effective approaches in science, technology, engineering, and mathematics.* Committee on Highly Successful Science Programs for K–12 Science Education. Board on Science Education and Board on Testing and Assessment, Division of Behavioral and Social Sciences and Education. Washington, DC: National Academies Press.

National Research Council. (2012). *Education for life and work: Developing transferable knowledge and skills in the 21st century.* Washington, DC: National Academy Press.

Pittsburgh Science of Learning Center. (2007). *Powerful mathematics instruction.* Retrieved from www.learnlab.org/research/wiki/images/f/ff/Accountable_T_Lit_Review.pdf

Provasnik, S., Kastberg, D., Ferraro, D., Lemanski, N., Roey, S., & Jenkins, F. (2012). *Highlights from TIMSS 2011: Mathematics and science achievement of U.S. fourth- and eighth-grade students in an international context* (NCES 2013-009). Washington, DC: National Center for Education Statistics, Institute of Education Sciences, U.S. Department of Education.

Reeves, D. (2006). *The learning leader: How to focus school improvement for better results.* Alexandria, VA: Association for Supervision and Curriculum Development.

Resnick, L., & Hall, M. (1998). Learning organizations for sustainable education reform. *Daedalus, Journal of the American Academy of Arts and Sciences, 127,* 89–118.

Rogers, E. (1995). *Diffusion of innovations.* New York, NY: Free Press.

Webb, N. (2002). *Depth-of-knowledge levels for four content areas.* Madison, WI: Wisconsin Center for Educational Research. Retrieved from wikipedia.org/wiki/100_metres#Youth_.28under_18.29_boys

Index

Abstract reasoning, 36–37, 62, 65
Achievement data, 122–123
Adding It Up (NRC), 15, 52
Adoption stages of change, 126–127
Advanced degree of proficiency, 59–60
Assessment
   examples, 11, 12 (box), 13
   formal, 47–48, 122
   inferences on, 136, 138
   informal, 47–48, 122
   rigor and, 19–20, 134
   tests, 8–9, 46–48
   See also Classroom visits; Formative
      assessment
Attention spans, 72–73

Balka, D., 74–76
Bloom's Taxonomy, 17
Boaler, J., 50, 72
Bob's story. *See* Principals
Brookhart, S., 48

CCS Content Standards and Practices, 22
CCSSM (Common Core State Standards
   for Mathematics), 20, 21
CESSM (Curriculum and Evaluation
   Standards for School
   Mathematics), 20
Challenges of teaching, 1, 10–11
Change
   adoption of, 104–105, 126–127
   classroom visits for, 105–106
   culture of, 106
   difficulty of, 150
   documenting, 163
   process of, 103–105
   supporting teachers in, 125–126, 162
   systemical, 147
Change Chain Reaction, 160–161
Classroom climate, 137, 138–139.
      See also Culture
Classroom visits
   Classroom Visit Tally, 93, 94 (figure)
   Classroom Visit Tally-Teachers Form,
      113 (figure), 114, 130
   Connecting Actions Chart, 107,
      107 (table), 108

importance of, 101
leaders making, 156
math coach scenario,
      98–99, 108–111
productive conversations
      about, 101–103
reasons for, 152
specified visits, 123–125
supporting change, 105–106
types, 92–93
Coaching for rigor, 151–154. *See also*
   Leadership
Collaboration
   classrooms with, 148–149
   culture of, 31
   isolation as obstacle to, 80
   making sense of problems with, 35
   Mathematics Collaborative
      Log, 119, 120 (figure)
   need for, 152
   rigor and, 26
   student activities for, 77–79
   team leadership and, 27
   See also Classroom visits; Learning
      communities
Common Core
   CCS Content Standards
      and Practices, 22
   CCSSM, 20, 21
   as exciting initiative, 27
   rigor and, 14
   united approach to, 1–2
   writing group, 33–34
   See also Standards for
      Mathematical Practice
Communication, 149, 153, 156
Conceptual development, 17
Conference Board of the
   Mathematical Sciences, 9
Confidentiality, 127, 161
Connections
   Connecting Actions Chart, 107,
      107 (table), 108
   curriculum requiring, 148
   meaningfulness of, 71–72
Construct viable arguments, 37–38
Consultants, 92–93

Content, 22, 70–71, 135, 138
Continual learning, culture of, 153
Contrasting lesson example, 23–26
Conversations, 101–103, 115, 161
Coteaching opportunities, 152
CRESST (National Center for
    Research on Evaluation, Standards,
    and Student Testing), 18
Critiques of others' reasoning, 38–39
Culture
    of change, 106
    of collaboration, 31
    of learning, 153, 157
    shifts in, 150
Current learning and formative
    assessments, 51–52
Curriculum, 148, 151–152, 155

Daggett, W., 17–18
Data
    analysis of, 102
    gathering assessments collectively,
        122–123
    tests, 8–9, 46–48
    See also Classroom visits
Deeper learning, 20, 132
Depth of Knowledge (DOK) scale, 18
Depth of mathematical
    understanding, 71–73
Differentiated instruction, 74–79
Diversity of students, 73–79
Documentation
    Linking Responsibilities Actions/
        Indicators, 163, 164 (figure)
    Teacher Implementation Progression,
        127, 128–129 (figure), 130
DOK (Depth of Knowledge) scale, 18

ELL (English Language Learners), 74–76
Engagement, indicators of, 54
Engaging Schools (NRC), 15
English Language Learners
    (ELL), 74–76
Euclidean geometry, 44
Evaluation of teachers, 9–10, 81–82, 91
Expectations, 156
Experimentation, 125–127
Explanations, asking for, 53

Feedback, 48, 101
Formal assessment, 47–48, 122
Formative assessment
    change occurring from, 103–105, 122
    current learning and, 51–52
    defining and refining, 48–51
    intervention and, 52–54, 53 box
    observation and listening for, 79
    synergy and, 56–57
Foundation definition, 7

Gavin, G., 73
Geometry example, 66
Graphing example, 66

Hall, G., 14, 103, 104–105
Hallinger, P., 81
Heck, R. H., 81
Heritage, M., 48
Hexagon example, 11–13, 12 box
Hord, S., 14, 103, 104–105
How People Learn (NRC), 15
How Students Learn (NRC), 15, 49
Hull, T., 74–76

Identified content teaching, 70–71
Indicators of engagement, 54
Informal assessment, 47–48, 122
Initial degree of proficiency, 59–60
Inputs, 148–149, 151–152, 155–156
Instructional research, 54–56, 55–56 box
Instructional strategies
    rigor in, 22
    sequence of, 61, 76–79, 168–169
    shifts in, 20
Instruction inferences, 135–136, 138
Intermediate degree of proficiency, 59–60
Intervention, 52–54, 53 box, 57
Isolation problem, 80
Issues
    about, 69
    depth of mathematical understanding,
        71–73
    identified content teaching, 70–71
    learning opportunities for diverse
        learners, 73–76
    Proficiency Matrix for diverse learners,
        78–79
    Strategy Sequence Chart for diverse
        learners, 76–79
    See also Obstacles to success

Leadership
    change and, 162
    Change Chain Reaction and, 160–161
    definitions, 27–28
    experimenting by, 126–127
    rigor and, 155–158
    role of, 151
    using achievement data, 122–123
    See also Principals
Leadership teams
    classroom visits and, 92
    communicating, 149
    defining roles/responsibilities, 27–31
    role of, 30
    student focus of, 83
Learning communities, 31
Linking Responsibilities Actions/Indicators,
    160–163, 164 (figure)

Listening strategies, 79, 156
Literacy and Numeracy Secretariat, 74

MAAT (Mathematical Adoption Analysis
    Tool), 83, 84–86 (figure), 87
Math coach scenario, 98–99, 108–111
Mathematical Adoption Analysis Tool
    (MAAT), 83, 84–86 (figure), 87
Mathematical reasoning skills, 36–37,
    43–44, 65
Mathematical relationship
    problem, 28–29
Mathematics Collaborative Log,
    119, 120 (figure)
Mathematics leader definition, 28
Meaningful connections, 71–72
Measurement examples, 41
Metacognitive realizations, 78
Missions, 149
Mistaken efforts, 82–83
Modeling process, 39–40, 65
Momentum, 159
Monitoring of students
    about, 90
    classroom visits for, 90–93, 94 (figure), 99
    evaluation and, 91
    math coach scenario, 98–99, 108–111
    student focus, 91–92
    teachers' self-assessment of students,
        93, 95–97 (figure), 98
Monitoring of teachers
    Classroom Visit Tally-Teachers
        Form, 113 (figure), 114
    conversations about data, 115
    importance of, 112
    individual needs, 115, 116–118 (figure), 119
    momentum and, 159
Moylan, K., 73
Multiple-choice tests, 10, 50–51
Murphy, J., 81

National Center for Research on
    Evaluation, Standards, and Student
    Testing (CRESST), 18
National Council of Teachers of
    Mathematics (NCTM), 49, 55
National Mathematics Advisory Panel, 49
National Research Council (NRC)
    Adding It Up, 15
    defining deeper learning, 132
    domains of competence, 20
    Engaging Schools, 15
    How People Learn, 15
    How Students Learn, 15, 49
    STEM Report, 46
    views on assessment, 49
    views on rigor, 15–16, 19
NCTM (National Council of Teachers of
    Mathematics), 49, 55

Observation of students. See Classroom
    visits
Obstacles to success
    about, 69
    evaluation of teachers, 9–10, 81–82, 91
    failure to monitor students, 82
    isolation problem, 80
    lack of knowledge about change, 104
    MAAT used to overcome, 83,
        84–86 (figure), 87
    mistaken efforts, 82–83
    systemic conditions, 90–91
    See also Issues
Ongoing formative assessments. See
    Formative assessment
Open-door policy, 90–91. See also
    Classroom visits
Outcomes, 149–150, 153–154, 156–157

Partnership for Assessment of Readiness
    for College and Careers (PARCC),
    9–10, 12 box, 18
Patterns, 43
Perseverance, 35–36
Planning process, 62–66, 63–64 (table)
Powerful Mathematics Instruction, 17, 18–19
Precision, 41–42
Principals
    forming steering committees, 29–30
    responsibilities, 29
    role of, 30–31
    story of, 31–32, 45, 57–58, 67
Problems, making sense of, 34–35
Productive conversations, 101–103, 161
Professional conversations, 161
Professionals on rigor, 16–19
Proficiency Matrix
    about, 3–4
    correlations with, 130
    critiquing reasoning, 65
    definitions, 22
    differentiated instruction example, 74–76
    formative assessment and, 62–66,
        63–64 (table)
    modeling, 65
    organization of, 59–61
    principal's story and, 45
    purpose of, 4
    sequencing and, 61, 78–79, 168–169
    Standards for Mathematical Practice
        and, 62, 133–134
    Standards of Student Practice, 166–167
    synergy and, 56–57
PSSM (Principles and Standards for School
    Mathematics), 20

Reasoning skills, 36–37, 43–44, 65
Recitation, 52
Reflection, 102

Regularity in repeated reasoning, 43–44
Research, 72–73, 133. *See also* National Research Council (NRC)
Reteaching process, 52
Reverse visits, 124–125
Rigor
    assessment and, 19–22, 47–48, 122
    categories of progress on, 134–139
    coaching for, 151–154
    contrasting lessons demonstrating, 23–26
    dictionary/thesaurus meanings, 16
    goal of, 14–15
    importance of, 164
    indicators of, 18–19
    NRC on, 15–16, 19
    outcome as, 137
    professionals on, 16–19
    Proficiency Matrix, 3–4
    relevance and, 17–18
    requirements of, 13
    shifts leading toward, 60–61
    Standards for Mathematical Practice and, 44–55, 133–134
    state departments of education on, 16
    steps for, 132–133
    teaching for, 148–150
    transforming classrooms for, 26
Rigor Analysis Focus Guide, 142, 143–145 (figure)
Rigor Analysis Form, 139, 140–141 (figure)
Rigor Comparison Chart, 21 (table)
Rigor/Relevance Framework, 17–18
Rogers, E., 104
Rounding example, 12 box

School leader definitions, 27
Sequence strategies, 61, 76–79, 168–169
Sharing process, 77–79
Skills, integrated approach to, 148
Smarter Balanced, 9–10, 18
Spontaneous evidence, 48
Standards for Mathematical Practice
    about, 22, 28, 33–34
    Connecting Actions Chart, 107, 107 (table), 108
    constructing viable arguments, 37–38
    critiquing others' reasoning, 38–39
    inferences from, 134–137
    list of, 2
    making sense of problems, 34–35
    modeling process, 39–40, 65
    perseverance, 35–36
    precision, attending to, 41–42
    principal's story and, 45
    Proficiency Matrix and, 62, 133–134
    reasoning skills, 36–37, 43–44, 65
    recognizing, 133
    regularity in repeated reasoning, 43–44

    rigor and, 44–45, 133–134
    strategic use of tools, 40–41
    structure of mathematics and, 42–43
    studying, 33
    support for, 9
    win-win situation from, 11
Standards of Student Practice, 166–167
State departments of education, 16
Steering committees, 29–30
Strategy Sequence Chart, 76–79
Structure of mathematics, 42–43
Students
    Change Chain Reaction, 160–161
    focusing on, 55–56, 83, 91–92
    monitoring, 82
    thinking/reasoning, 49–50, 50–52
    See also Classroom visits
Summative assessments, 47–48
Symbolic language, 41
Synergy, 56–57

Teacher Implementation Progression (TIP), 127, 128–129 (figure), 130
Teacher Planning Guide, 119, 121 (figure)
Teachers
    Change Chain Reaction, 160–161
    content knowledge, 17
    evaluation of, 9–10, 81–82, 91
    intervention techniques, 105–106
    job outcomes, 154
    learning shifts, 10
    requesting classroom visits, 125
    self-assessing students, 93, 95–97 (figure), 98
    teaching for rigor, 148–150
Teacher Self-Assessment for Personal Actions, 115, 116–118 (figure), 119
Technology advances, 2–3
Tests, 8–9, 46–48. *See also* Assessment
Themes for promotion of rigor, 19
Trends in Mathematics and Science Study (TIMSS), 46
TIP (Teacher Implementation Progression), 127, 128–129 (figure), 130
Tools, strategic use of, 40–41
Trends, 2–3
Trust, 161

Urgency, sense of, 7

Validity tests, 124
Viable arguments, 37–38
Visible learning, 51–52
Visible thinking, 57
Visible Thinking Activities (Balka and Hull), 74–76
Vision, 149

Webb, N., 18